The cost-benefit to the NHS arising from preventative housing interventions

Helen Garrett, Mike Roys, Selina Burris and Simon Nicol

The research and writing for this publication has been funded by BRE Trust, the largest UK charity dedicated specifically to research and education in the built environment. BRE Trust uses the profits made by its trading companies to fund new research and education programmes that advance knowledge, innovation and communication for public benefit.

BRE Trust is a company limited by guarantee, registered in England and Wales (no. 3282856) and registered as a charity in England (no. 1092193) and in Scotland (no. SC039320). Registered office: Bucknalls Lane, Garston, Watford, Herts WD25 9XX
Tel: +44 (0) 333 321 8811
Email: secretary@bretrust.co.uk
www.bretrust.org.uk

IHS (NYSE: IHS) is the leading source of information, insight and analytics in critical areas that shape today's business landscape. Businesses and governments in more than 165 countries around the globe rely on the comprehensive content, expert independent analysis and flexible delivery methods of IHS to make high-impact decisions and develop strategies with speed and confidence. IHS is the exclusive publisher of BRE publications.

IHS Global Ltd is a private limited company registered in England and Wales (no. 00788737).
Registered office: The Capitol Building, Oldbury, Bracknell, Berkshire RG12 8FZ. www.ihs.com

BRE publications are available from www.brebookshop.com
or
IHS BRE Press
The Capitol Building
Oldbury
Bracknell
Berkshire RG12 8FZ
Tel: +44 (0) 1344 328038
Fax: +44 (0) 1344 328005
Email: brepress@ihs.com

Printed using FSC or PEFC material from sustainable forests.

FB 82
First published 2016
ISBN 978-1-84806-443-0

Acknowledgement

The authors would like to thank Martin Hodges (Care & Repair England) Doug Stem (Foundations) and Anne Kane (Habinteg Housing Association) for their help in sourcing photographs and case studies for this publication.

Cover images:
Top right: © WE Care & Repair England
Bottom right: © Habinteg Housing Association

Contents

Foreword

To anyone involved in adapting disabled people's homes, the enormous benefits in terms of individuals' quality of life are self-evident, as are the impacts on reduced social care need and the extension of independent living.

However, in light of the major changes taking place in the NHS, and the drive to integrate health and care, there is a growing demand for research which quantifies home adaptations' impact on health and care costs.

Home adaptations are multifaceted. They can range from installation of a small grab rail to building a tailored home extension. Likewise, disabled people are a heterogeneous group with a wide range of housing adaptation requirements, and so no single study can ever provide a definitive figure for all of the cost-benefits of all home adaptations.

The last definitive home adaptations evidence summary, *Better outcomes, lower costs** concluded that while not all adaptations save money (some 'just' improve quality of life) 'where they are an alternative to residential care, or prevent hip fractures or speed hospital discharge; where they relieve the burden of carers or improve the mental health of a whole household, they will save money, sometimes on a massive scale'.

Studies of aspects of home adaptations have demonstrated the importance of an adapted home to well-being, independence and health, with some identifying cost-benefits, particularly to social care. What has been less studied is the preventative nature of home adaptations. Quantifying the potential cost-benefits of particular home adaptations specifically to the NHS, for example reducing the risk of an acute incident which results in greater health care need, is therefore a high priority.

The Department of Health is now the main provider of state funding for home adaptations, with the budgets of Disabled Facilities Grants now incorporated into Better Care Funds, which in turn are managed by Health and Wellbeing Boards. While there is broad acknowledgement of the preventative role of adaptations, for example in reducing falls, those faced with difficult funding decisions are looking for hard evidence of the cost-benefits of particular interventions.

It is notable that the *NHS Five year forward view*[†] called for both a 'radical upgrade in prevention' and also for expansion of evidence-based action. This BRE report is timely as it takes a new approach to modelling the medium to longer term potential cost-benefits to the NHS of pro-active modification of hazardous homes lived in by people with long-term health conditions and/or disability.

BRE is at the forefront of quantifying the costs to the NHS of poor housing. In this new research reported in this publication, BRE demonstrates again how its approach to modelling and data analysis can contribute to an evidence base around the cost-benefits of prevention, demonstrating the potential role of pro-active housing interventions in reducing NHS costs.

Care & Repair England, and other members of the national Home Adaptations Consortium, will continue to champion the key role of home adaptations in improving the lives of disabled people and we welcome this contribution to the evidence base that can help to support those efforts.

Sue Adams OBE, CEO

Care & Repair England and Chair of the
Home Adaptations Consortium

---

* Heywood F and Turner L (2007). *Better outcomes, lower costs: implications for health and social care budgets of investment in housing adaptations, improvements and equipment. A review of the evidence.* London, Office for Disability Issues.

† NHS England (2014). *NHS Five year forward view.* London, NHS England. www.england.nhs.uk/wp-content/uploads/2014/10/5yfv-web.pdf.

Executive summary

In 2014 a 'Bletchley Day' workshop organised by Care & Repair England, was tasked with considering ways to demonstrate the investment value of home adaptations and modifications through the production of better evidence. Previous health cost-benefit assessments of home adaptations have largely examined these for individual household scenarios. BRE Trust agreed to fund the research, which attempts to model the cost-benefit of some common preventative home interventions on a larger scale using national data sources. It is important to stress from the outset that this research was not designed to demonstrate the economic benefits to the state of Disabled Facilities Grants (DFGs) at the national level.

Unlike previous research into home adaptations, the main aim of this BRE research project was to consider the cost-benefits of preventative home interventions by reducing the need for NHS treatment and reducing the subsequent need for reactive home adaptations in as many cases as possible. The NHS treatment and adaptations relate to households with known health problems and who are living in homes with serious hazards assessed through the Housing Health and Safety Rating System (HHSRS). This reduction in the need for home adaptations that result from a preventable health problem allows the DFG budget to provide help to those households where preventative action is not applicable. The cost-benefits of DFG-funded adaptations to the NHS and social care, particularly when compared to the cost of residential care (highlighted in the literature review in the Appendix) continue wherever such adaptations are carried out.

This research discussed in this report uses the same basic methodology developed to calculate the data published in *The real cost of poor housing* (Roys et al, 2010) and summarised in BRE Information Paper 16/10 *Quantifying the cost of poor housing* (Nicol et al, 2010). Using HHSRS information from the English Housing Survey (EHS) on the risk of a home incident occurring and its likely impact on health, combined with information from the NHS on treatment costs, BRE research estimated that it was costing the NHS some £600 million per annum, in first year treatment costs, to leave people living in the poorest housing in England (a home with at least one of the most serious hazards). Following improvements to the modelling, including a broader definition of poorer housing to include all substandard homes, this estimate has been revised to £1.4 billion (Nicol et al, 2015).

For this research, households containing someone with a long-term sickness or disability were identified as the most appropriate group for the new national model, namely the group of households most likely to be in need of some form of home intervention owing to their physical and medical circumstances. In 2012, the EHS estimates that this group comprised 6.4 million of all English households. Of these, around 854,000 lived in a home with at least one Category 1 HHSRS hazard. Furthermore, around 2.2 million of these households lived in homes with less serious hazards but which presented a higher than average risk of harm. In total, therefore, around 3 million of these households had significantly higher than average risks of a harmful event occurring within the next 12 months.

A number of assumptions have been tested during the costing exercise, but our best estimate suggests that leaving long-term sick and disabled occupants in homes with significant hazards is costing the NHS nearly £414 million per annum in first year treatment costs alone. Furthermore, if we add the costs of installing a potentially more costly home adaptation following a harmful event, such as a fall on stairs, because remedial action has not been undertaken, the economic cost rises to around £529 million per annum. The potential savings to the DFG budget are important given the increasing need for home adaptations as a result of, for example, our ageing society, and the pressures on public expenditure.

The largest NHS costs occur due to the treatment of harms arising from excess cold. Although excess cold comprises 8% of all the hazards identified among the homes of long-term sick and disabled households, it comprises 34% of the £529 million cost to the NHS identified in this research. In addition, lack of remedial action to address the risk of falls, particularly those associated with stairs, incurs notably higher NHS treatment costs. Falls on stairs comprise 38% of Category 1 hazards and 22% of other significantly worse than average hazards and comprise around a quarter (24%) of the NHS costs identified.

The total cost of remedial works to mitigate the risk associated with these hazards in homes occupied by someone at risk of harm is estimated to be £6.4 billion. Although a huge cost, the expenditure should benefit people in all 3 million houses to which it is applied. Furthermore, the average cost of work per household is just £2,130.

On average the payback period to the NHS to mitigate each type of harm through a large-scale preventative programme of targeted interventions in homes with Category 1 hazards is 15 years. The best paybacks come from mitigating falls on the level (5.2 years), falls on stairs (5.9 years), falls in baths (6.5 years) and excess cold (6.9 years). It is important to note that there are potentially additional savings resulting from such a preventative approach, including savings to social care. It is also important to note that these payback periods are not intended to represent the timescale of any benefit to the individual household. Indeed the household will receive the benefits of a home intervention, such as a reduction in risk of harm and a likely improvement in the quality of life, at any time after the intervention. These benefits are demonstrated through the case studies contained in this report.

The estimated incidence of prevented DFG demand in the modelling is likely to be small compared to the number of households who need adaptations through a DFG, most of whom live in homes without a serious hazard and require an adaptation in order to carry out activities of daily living, for example wash, dress and prepare food. This research shows that over half of the 6.4 million long-term sick and disabled households live in a home where the risk of serious harm, as assessed through an HHSRS assessment, does not exceed the national average. Nonetheless a significant proportion of these households will still require a home adaptation due to their difficulties in maintaining independence on a daily basis because of their physical and medical circumstances.

As previous BRE research for DCLG cited (BRE for DCLG, 2011), while an estimated 1 million households require adaptations to their home, there is no robust and definitive means to establish the potential demand for DFGs in the future, let alone the scale and cost of adaptations paid for by households themselves or other via charitable sources.

Some harmful events in the homes may result in either the introduction or extension of home-care, but the new national model is unable to assess the cost-benefit of home interventions to social care. Heywood and Turner's review (2007) into the benefits of investment into home adaptations highlights that savings from adaptations can vary from £1,000 to £29,000 per annum depending on the level of care needed. Falls in the home may also precipitate a move into long-term care for older people. In the case of young children or younger adults, there may be other economic costs to both the individual and the state resulting from an injury at home, for example loss of potential earnings or loss of income due to absence from work and subsequent loss of taxation revenue or increased need for state benefits. Consequently, although not quantifiable, the preventative home interventions advocated in the report will have a positive impact in reducing social care costs and other societal costs.

This research has, therefore, demonstrated some of the potential cost-benefit to the NHS of undertaking preventative, pro-active home interventions for households with a long-term sickness or disability, where the risk of accidents in their home are significantly worse than the national average. Furthermore, it has been possible to demonstrate how the cost of this preventative action is partially offset through subsequent savings to the DFG budget, so providing an additional payback to the state and society for the preventative work. It is hoped that this research will enable a more informed case for investment in preventative housing interventions and adaptations. These improve people's health and make sound economic sense, as well as saving public money in the longer term. Furthermore, it is important to recognise that many important benefits of home interventions are associated with an improvement in people's quality of life, such as feelings of dignity and independence. The case studies used in this research will demonstrate the importance of this outcome.

It may be possible to adapt or enhance the methodology used for the report so that it can be developed into a practical tool to enable local housing and health providers to demonstrate the value of all forms of preventative housing interventions where there are perceived risks to the safety of people in their homes. There are several issues that would need to be considered for such a practical tool including:

- the current uncertainty in the estimate of the total number of adaptations being undertaken (from all sources) and the average cost of these adaptations
- the feasibility of 'creating' a single method of assessing the need for home adaptations that can be applied by professionals involved in assessing risks in the home and the mitigations that could reduce that risk as well as making it easier for people to live safely and independently at home
- how the need for adaptations arose; if the need was due to previous harm, ie a fall resulting in referral for adaptation from a GP or from a hospital, what the total cost of that harm was, and who covered this cost.

It is evident, however, that more research is required into the economic benefits of home adaptations and other interventions, particularly into the potential wider costs savings to NHS/social care budgets so that these are better understood.

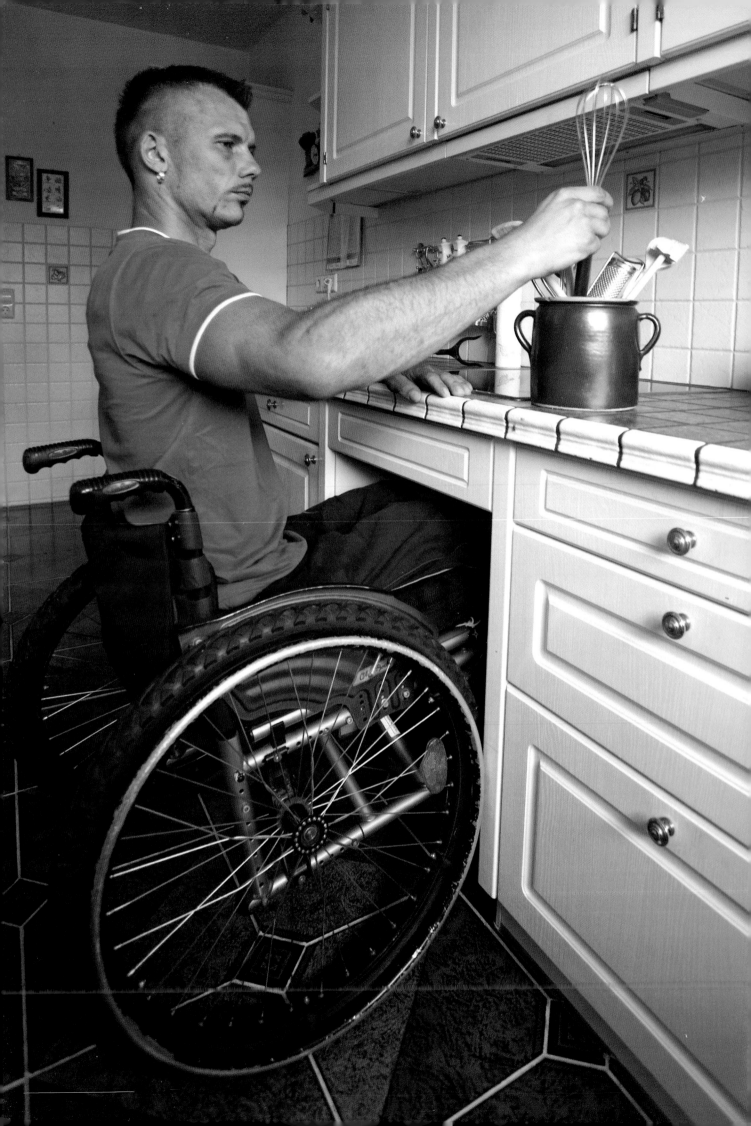

1 Introduction

In 2014 a Bletchley Day workshop organised by Care & Repair England, was tasked with considering ways to demonstrate the investment value of home adaptations and modifications through the production of better evidence. One outcome of this workshop was a request for BRE Trust to provide research that contributed to this evidence base.

Building on the methodology of previous BRE Trust research into the costs of poor housing, the key objectives of this research were to:

- provide a general overview of the support available for disabled and older people
- review existing research that has attempted to measure the cost-benefit of home interventions
- estimate the nature and quantity of hazards that exist in the homes of households who are most at risk of harm from these hazards
- estimate the reduction in cost to the NHS which would arise from undertaking remedial work/home modifications to mitigate the risks of these hazards
- estimate to what extent the costs of mitigation work can be offset by savings to the Disabled Facilities Grant (DFG) adaptations budget
- consider some additional costs to society of not undertaking remedial work/home modifications in the homes of those who have the potential to benefit from this action
- provide further empirical evidence of the benefits of home interventions through case studies.

Like its predecessors, this research aims to demonstrate that enabling people to live safely in their home makes economic sense, by reducing NHS expenditure, as well as improving the quality of lives of the people who benefit from them. In making the case for government investment in home interventions to keep people in their own homes, it is hoped that the research will also help local authorities and charities with limited resources to target funding where it offers the best value.

It is hoped that the research will be of particular interest to all government departments responsible for health and housing, Public Health England, the NHS, local authorities, social housing providers, Age UK, Home Improvement Agencies, the Chartered Institute of Housing and the Chartered Institute of Environmental Health and students of public health and housing.

Further information on the research methodology underpinning this research will be available in a forthcoming BRE Trust publication, *The full cost of poor housing* (FB 81).

2 Research background

Home adaptations that enable homes to become safer and promote independent living have the potential to transform the quality of life for those who need them. They also help deliver some government health and social care key policy objectives, for example, prevention of unnecessary hospital stays, reducing strain on carers, and promoting the social inclusion of people who may otherwise have to remain at home or live in just a few rooms of their home. Demand for home adaptations is increasing nationally due to our ageing society and medical advances. Innovations in the care of pre-term babies and in medical support for disabled children have, for example, led to an increasing number of children and young people needing home adaptations. Around 40,000 people every year (mainly older people), benefit from home adaptations provided through a DFG administered by local authorities. These grants, however, represent a small proportion of those who undertake adaptations to their home or who would benefit from this form of assistance.

Until recently, home adaptations including DFGs, and repair work to homes to make them safer, have been traditionally associated more strongly with 'housing' policy and provision rather than with 'health'. The impact of inadequate and poor housing upon health and well-being is, however, being increasingly recognised in joint housing and health policy and delivery. Furthermore, there is a growing body of evidence (as shown by the literature review in the Appendix) that demonstrates that the home interventions can have significant benefits for health and social care expenditure.

In 2014 a Bletchley Day workshop, organised by Care & Repair England, was tasked with considering ways to demonstrate the investment value of home adaptations and modifications through the production of better evidence. This evidence is required for many interrelated reasons, including:

- **The challenges faced by our ageing population**
  Office for National Statistics (ONS) predictions suggest that the number of people of pensionable age in the UK will increase from 12.3 million in 2012 to 16.1 million in 2037. By mid-2037, 1 in 12 of the UK population will be aged over 80 years (Office for National Statistics, 2013). The vast majority of older people currently live in ordinary homes (rather than specialist housing) and it is likely that our ageing population will continue to have a preference for independent living arrangements supported by community-based aged care services. There is an aspiration that, in remaining at home, it is safe, affordable and warm; that is, the home environment itself is not disabling.
- **Increased pressure on NHS and social care resources**
  Prevention of the need for NHS resources is critical to the Care Act 2014 statutory guidance (DoH, 2014). Home adaptations play a key part in improving indicators in the Public Health Outcomes Framework (DoH, 2012) by reducing hip fractures in people over 65 years of age, reducing emergency readmissions within 30 days of hospital discharge and reducing excess winter deaths. They also contribute to the Adult Social Care Outcomes Framework (DoH, 2014) by, for example, enhancing the quality of life for people with care and support needs, and delaying and reducing the need for care and support.

- **The integration of funding arrangements for DFGs into the Better Care Fund (BCF)**
  The June 2013 Spending Round announced the creation of a £3.8 billion BCF, pooled budget for local integrated health and social care services, based on agreed strategies between the NHS and local authorities. This new fund included all central funding for DFGs (£220 million) in 2015/16. In 2016/17, the Better Care Fund will be increased to a mandated minimum of £3.9 billion with the national allocation of funding for DFGs set to almost double to £394 million (Department of Health and DCLG, 2016).

Previous BRE Trust publications *The real cost of poor housing* (Roys et al, 2010) and *Quantifying the health benefits of the Decent Homes programme* (Garrett et al, 2014) have demonstrated how investment in housing can save the NHS (and England as a whole) money in the long term. One outcome of the Bletchley Day workshop was a request for the BRE Trust to produce a further report in the cost of poor housing series on the health cost-benefit of preventative housing adaptations and modifications. This publication is the outcome of that request.

Interestingly, a House of Commons briefing paper has also advocated the need for research in this field:

> '*We need to compile compelling evidence to demonstrate how money spent on adaptations will save money on health and care costs. This needs to take the form of theoretical cost benefit analysis, possibly using a similar approach used to that developed by BRE in recent work on the costs of poor housing (Roys et al 2010), and case studies to give concrete examples of how this works in practice.*'
>
> Wilson, 2013

Previous research on the health cost-benefits of home adaptations and modifications, particularly, DFGs, has largely examined these for individual household scenarios. The BRE research project, reported in this publication, attempts to model the cost-benefit of some home interventions to reduce the risk of harms caused by hazards in the home on a larger scale using national data sources.

It differs from other research as it suggests preventing the need for adaptations (for example through DFGs) in as many homes as possible. This would then enable the monies available for adaptations to be spread among those households where intervention was not pre-taken because the household lived in a home without a significantly higher than average risk of harm.

2.1 Supporting people in their homes

Some people, of all ages, who have physical or sensory impairments, do not live in a home that is suitable for their needs. Also, as people get older, their housing needs often change and their homes are no longer suitable without some form of modification.

The former government's housing strategy *Laying the foundations: a housing strategy for England* (DCLG, 2011) outlines policies which seek to provide a better deal for older people, with greater choice and support to help them live independently. The housing strategy outlines the following initiatives to enable this:

- investment into the FirstStop information and advice service[a]
- continued help with small repairs, security and safety in the home through funding for handyperson schemes (2011/15) to deliver small home repairs and adaptations
- promotion of Lifetime Homes standards in new homes to enable accessibility and adaptability
- help through Home Improvement Agencies (HIAs) which assist around 250,000 older and disabled people.

HIAs[b], normally known as Care & Repair England[c] or Staying Put, help elderly, disabled and other vulnerable people to achieve or maintain independent living and enable them to live more comfortably and safely in their own homes[d]. HIAs deliver around half of all DFG-funded adaptations on behalf of local authorities and also deliver adaptations funded directly by housing association landlords. Other HIA[e] services include:

- housing advice, including help with moves to more suitable accommodation when staying put is not the best or feasible option. (In some cases disabled people are helped through the process of purchasing or renting a more accessible property, or helped through relocation grants and loans.)
- handyperson services, including small home repairs, home safety and security checks and adaptations
- hospital discharge services, for example equipment delivery, moving beds/chairs
- falls and accident prevention checks and remedial action
- energy efficiency advice and remedial action
- installation of assistive technology.

Together home improvement and handyperson services provide support for over a quarter of a million people every year, underpinning national policies to increase choice and empower disabled and other vulnerable people (Figure 1). The national evaluation of handyperson services demonstrated that its services can be cost beneficial with the programme estimated to have outweighed the costs of provision by 13%. The biggest costs that can be avoided relate to social care (DCLG, 2012).

Figure 1: Home independence centres in Bristol, Weston-super-Mare and Yate which provide members of the public with examples of available adaptive technologies. Images © WE Care & Repair

[a] FirstStop is a free, independent service for older people, their families and carers and aims to help older people make informed decisions about their housing, care and support options and to help them maintain independent living in later life.

[b] Foundations is the current national body for HIAs. See www.foundations.uk.com/home.

[c] See http://careandrepair-england.org.uk.

[d] The majority of HIA services are operated by housing associations; some are provided 'in-house' by local authorities, while others are small independent organisations, usually with a charitable status.

[e] Includes the work of the Foundations Independent Living Trust and its Warm Housing Scheme. See www.filt.org.uk/news-items/warmat-home/#.VbtBxTZwZdg.

The Supporting People programme is a government programme for the funding, planning and monitoring of housing-related support services, helping vulnerable people to live as independently as possible in the community.

Examples of the type of support available include:

- a warden or scheme manager
- a community alarm service
- general counselling and support, including befriending services
- help to deal with claims, social security benefits and managing money
- advice or assistance with shopping and errands
- assistance to engage with individuals, professionals and other bodies.

In the 2010 Spending Review, the government announced that the Supporting People national funding levels would fall from £1.64 billion in 2010/11 to £1.59 billion in 2014/15. Furthermore the grant was no longer ring-fenced from 2009, allowing local authorities to spend their Supporting People allocation as they deemed appropriate.

Research commissioned by DCLG on the value for money provided by Supporting People funding concluded that the net financial benefits of the programme were £3.41 billion (Jarrett, 2012).

Some of the above-mentioned services and additional support are provided by organisations such as The Children's Trust[f], Age UK[g] and the Papworth Trust[h]. The Thomas Pocklington Trust[i] deliver a range of housing and support services for people with sight loss and also funds research aimed at improving and developing services, and promoting independence and quality of life, for example through appropriate design of domestic environments. It is not possible to list all of the organisations that provide help and support, but the role of older people's groups and the wider community and families must also be acknowledged for their work supporting disabled and vulnerable people at home.

2.2   Existing technologies

'The purpose of an adaptation is to modify the home environment in order to restore or enable independent living, privacy, confidence and dignity for individuals and their families. The focus is therefore on identifying and implementing an individualised solution to enable a person living within a disabling home environment to use their home more effectively rather than on the physical adaptation itself.'

Home Adaptations Consortium, 2013

Housing-related support through equipment and adaptations, often provided in partnership with other care services, can improve health and reduce demand for more costly health and social care services. This may enable the full benefits of other forms of care, such as family care, to be realised. This section details the various technologies available that enable people to remain living in their homes, where this is preferred.

The most common existing technologies available (Figure 2) are:

- portable aids: wheelchairs/frames/toilet seats/crutches or sticks/profiling beds/recliner chairs
- external ramps and grab rails
- entry phones
- front door replacements or widening and flush thresholds
- internal grab rails
- WC and bathing adaptations or more minor alterations: toilet seat/bath chair/hoist/cradles/low level bath/shower over bath
- WC and bathing adaptations: redesign or provision of showers (graduated shower floor) or wet room. Relocation of WC
- stairlift
- kitchen alterations: minor or redesign
- door widening/sliding doors (internal)
- home extensions/pod
- lighting, for example in stairwells/external light fittings
- telecare (see Section 2.4).

2.3 Case study: Handyperson services

Case study: Foundations

The use of handyperson services is demonstrated here through a case study provided through Foundations (www.foundations.uk.com).

Mrs G (84) owns her property, lives alone and is in receipt of means tested benefits with minimal savings. Mrs G had seen a council advert that detailed information about a Winter Home Healthy People (WHHP) scheme and a winter home check service and contacted her local HIA. She was in a desperate state as her emersion heater had sprung a leak and water would potentially damage the flat below her. Mrs G was emptying buckets into her bath, a problem that was exasperated by the fact that, due to her arthritis, she could not turn her mains stopcock off to stop the tank filling up constantly.

The HIA team arranged for a heating and plumbing engineer to visit within 3 hours of receiving Mrs G's call for assistance. While onsite, the engineer discovered that a new header tank would also need to be installed. The estimate was submitted the next day and additional funding was gained to ensure Mrs G was not left without hot water for any length of time. Foundations was also able to provide additional funding and Mrs G insisted on paying £200 toward the repairs, the remainder was paid for through the Winter Home Check scheme. The HIA also undertook a WHHP check and found several ways to support Mrs G in the interim while the more major repairs were organised. The HIA repaired her kitchen tap, supplied heaters and snug pack to keep her warm and undertook draught proofing to reduce cold air flow.

[f]   www.thechildrenstrust.org.uk.
[g]   www.ageuk.org.uk.
[h]   www.papworthtrust.org.uk.
[i]   www.pocklington-trust.org.uk.

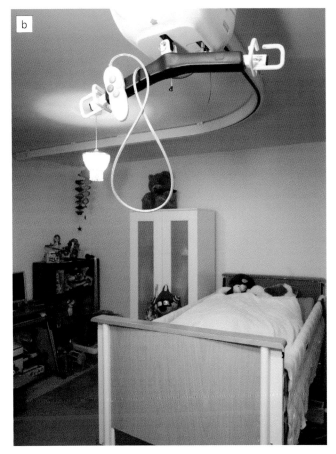

Figure 2: (a) and (b) show adaptive technologies for the bathroom and bedroom. (c) and (d) show adaptations to create level-access and a flush threshold for two homes of wheelchair users. Images © Habinteg Housing Association

telecare, digital participation services[j], and wellness services[k], are increasingly being offered to individuals in need of social care support as a way of helping them to remain independent and to promote quality of life for them and their families.

Telecare[l] is defined by the DoH as a service that uses:

> 'a combination of alarms, sensors and other equipment to help people live independently. This is done by monitoring activity changes over time and will raise a call for help in emergency situations, such as a fall, fire or a flood'.

Davies and Newman, 2011

2.4 Assisted living and telecare

The government considers technology and innovation to be key enablers for efficiently enhancing quality in services. This includes personalised and integrated assistive technology services, which support independent living in the community, and which prevent or reduce admissions to hospitals and care homes. Assisted Living Technology (ALT), provided through

It is estimated that between 1.6 to 1.7 million older people in England benefit from some form of telecare services (Clark, 2010). The use of community alarms to contact control centres normally forms the basis for the introduction of telecare measures. These include fall alarms, safety sensors for gas leaks and bath floods and 'wander' monitors for people with dementia.

[j]   Designed to educate, entertain and stimulate social interaction for people in need of social support.

[k]   Designed to encourage people to adopt and maintain a healthy lifestyle and to prevent or delay the need for support.

[l]   Telecare differs from Telehealth. Telehealth involves the remote monitoring of physiological data, such as blood pressure, pulse and weight, enabling consultation between health professionals or between a health professional and a patient by, for example, telephone.

Telecare has the capacity to:

- grow in provision as new problems emerge with daily living
- improve an older person's sense of security and self-confidence
- relieve some of the burdens and stress for carers
- be cost-effective for NHS and social care budgets by, for example avoiding or deferring an elderly person's move into residential care or a hospital.

In 2008, the DoH launched three pilot schemes (in Newham, Kent and Cornwall), known as the Whole System Demonstrators, to test the benefits of integrated health and social care supported by assistive technologies like telecare and telehealth (DoH, 2013). Each of the three pilot areas made their own decisions on the types of telecare equipment to be used to meet local assessments of need. On-going evaluation of the data from the pilot schemes is being undertaken by six major academic institutions. The Appendix summarises the results of the literature review on telecare services undertaken for this report. Further information on telecare can be found via http://telecareaware.com/telecare-aware-about.

2.5 Disabled Facilities Grants (DFGs)

DFGs fund essential home adaptations giving disabled people better freedom of movement into and around their homes, as well as providing access to essential facilities such as showers and toilets. DFGs can also be used to make homes safer through, for example improved lighting and heating. This is currently a mandatory grant administered by local authorities and may be subject to means testing, as well as, most often, a consultation with an occupational therapist who recommends the appropriate adaptations. Professionals working in the field of DFGs have long extolled the benefits of adaptations for those receiving them, such as improved mental health and quality of life, and a reduction in the use of more costly hospital or residential care. The case studies in this report demonstrate these benefits.

Funding for home adaptations comes from two main routes administered by local authorities:

- Community Equipment Services funding which pays for small items and minor home adaptations up to a value of £1,000, such as walking frames, bath seats, and grab rails. These are provided free of charge to the client.
- DFGs of up to £30,000. These mandatory grants are normally issued subject to a means test. Where an application is for a disabled child (or a qualifying young person), there is no means test. The average cost of a DFG grant is around £6,500.

When DFGs were introduced under the 1996 Housing Grants, Construction and Regeneration Act, central government provided 60% of total DFG funding, and the remainder was provided from local authority budgets. From 2008/9, however, local authorities were no longer required to match this funding. The current contributions from local authorities now vary considerably across the sector (Astral Advisory, 2013) and there is a long-term decline in local contributions as local authorities are having to rationalise expenditure.

The vast majority of DFGs are delivered to private sector households, predominantly owner occupiers, but a significant proportion of funding goes to housing association tenants. This is likely to increase as housing association budgets come under increasing pressure. While local authority tenants can, in theory, seek home adaptations via a DFG, local authorities normally deliver needed adaptations using their own housing resources. In addition, many local authorities spend a proportion of their DFG budget adapting homes owned by housing associations.

While variations in funding arrangements for home adaptations in the social sector are not the focus of this research, it is important that potential strains on existing budgets are recognised, particularly in the current economic climate. These financial constraints underpin the need for research to examine the value for money provided by home adaptations and, as this research does, put the case for preventative action in order to help alleviate the pressures on the DFG budget.

Types of work covered by mandatory DFGs include:

- widening doors and installing ramps and grab rails to make it easier to get into, and out of, the home
- providing or improving access to the WC, washbasin and bath (and/or shower) facilities
- improving or providing a suitable heating system in the home
- improving access and movement around the home
- adapting heating or lighting controls to make them easier to use
- improved lighting to enhance visibility
- ensuring the safety of the disabled person by, for example providing a specially adapted room in which it would be safe to leave that person unattended.

DFGs help around 40,000 people with mobility problems every year (Adams, 2015), mainly older people. This, however, represents a small proportion of those who need assistance. Also, as BRE research for DCLG cited (BRE for DCLG, 2011), while an estimated 1 million households require adaptations to their home, there is no robust and definitive means to establish the potential demand for DFGs in the future. Analysis using English House Condition Survey (EHCS) data for that research indicated that the total amount required to cover grants for all of those who were theoretically eligible under the existing rules was £1.9 billion at 2005 prices. This was more than 10 times higher than the total amount of DFG allocated in England in 2009/10 (£157 million).

Concerns have been expressed regarding the lack of speed and efficiency and cost-effectiveness regarding the DFG process (for example see Bristol University for the ODPM, 2005; Foundations, 2010; Papworth Trust, 2012; Age UK, 2014). Prompt assessment and delivery of adaptations is often essential for people to enjoy the security and comfort of their own home rather than being forced into residential or hospital care. This also enables the economic benefits of adaptations to be fully realised. While it was beyond the remit of the BRE research project to evaluate mechanisms of DFG delivery, and how these should be integrated with other housing support, the research aims to demonstrate the economic benefits to the NHS of prompt installation of home interventions to help alleviate pressures on DFG delivery.

2.6 Case studies: DFGs

Case study 1: Care & Repair England

Mr S, an 83-year-old widower, had lived in his two-bedroom bungalow for 22 years. Because of his limited mobility and falls in the home, he was provided with a care package that involved assistance with washing, shopping and cleaning the home. Mr S was unable to use the bath. Seeking help from his local authority, Mr S was sent the DFG enquiry forms but, due to ill health, was unable to complete them. West of England Care & Repair were allocated the case by North Somerset Council to provide the necessary help and discuss a number of housing options with Mr S, including moving to an extra-care sheltered housing scheme.

As Mr S wanted to remain in his home, a DFG was pursued and bathroom modifications including replacement of the bath with a level-access shower, seat and rails were undertaken (Figure 3). The works cost £4,256 (plus an HIA fee). Mr S chose to make a modest contribution to fully tile the bathroom to suit his personal taste.

The HIA also noted a risk of falls from trailing electrical cables and raised money from a charitable source to provide a wall-mounted electric fire. They also identified uneven and dangerous floor surfaces in the conservatory resulting from very poor workmanship by a previous local contractor. Remedial works via the agency's handyperson service removed this hazard.

- **Care costs:** The element of the care package relating to assistance with washing and bathing has been removed as Mr S is confident enough to use the facilities independently; a saving to the local social care budget.
- **Health costs:** No falls in the home have occurred since the works were completed, so reducing necessary contact with his GP and other health practitioners.
- **Dignity and confidence:** According to Mr S there has undoubtedly been an improvement in his personal dignity and improved mental wellbeing as a direct consequence of the work. He is feeling more positive about his quality of life and is exploring ways he can participate in local community activities to address the social isolation and loneliness he had been feeling.

Case study 2: Collaborative work by First Stop and Care & Repair England

Mrs M is 75 and lives alone, having been widowed 5 years ago. She has no family living locally. Mrs M is the owner occupier of a three-bedroom property with a first floor bathroom. She has several long-standing health problems including COPD, bronchiolitis, arthritis, spondylitis and heart problems. She has no savings.

Mrs M had been admitted to hospital with a severe chest infection. As a result of Age UK West Cumbria's involvement with discharge planning arrangements, a service was put in place to provide support, for example shopping and light housework tasks, when she was discharged. At Mrs M's request a further referral was made to Age UK West Cumbria's Help at Home Service as she realised she needed additional support to continue to live independently at home. She was going up her stairs on her hands and knees and coming back down seated on the stairs and was at risk of falling, particularly given her health problems. In addition she was now finding bathing very difficult and this was made worse by her deteriorating health. Mrs M

Figure 3: Home adaptations for Mr S. Images © Care & Repair England

did not want to consider moving as she has lived in the house for over 40 years and it was where she had raised her family. The Help at Home co-ordinator referred Mrs M to the Housing Options Service.

The Housing Options adviser (who is a trusted assessor for minor adaptations) visited Mrs M at home. Following a discussion about her housing options, Mrs M was referred to Adult Social Care for a DFG assessment. As a result of the occupational therapist assessment, a grant was recommended to install a stairlift and provide a fully accessible bathroom. The Housing Options adviser helped Mrs M to complete the application forms and also arranged with the local HIA for estimates to be obtained.

Mrs M's application was approved. The Housing Options adviser liaised with the local HIA to ensure that Mrs M could cope while the work (which took 6 weeks) was carried out. Mrs M was delighted. She had not known about any options to enable her to remain in her own home and had been concerned that with her limited income her only option would have been to move into residential care.

2.7   The accessibility of the English housing stock

Before reviewing the evidence regarding the benefits of home adaptations, English Housing Survey (EHS) data further demonstrates the potential future demands for adaptations by considering the accessibility of the English housing stock.

Most of the current housing stock does not benefit from a combination of key accessibility features which allow for changing needs without some form of adaptation. The EHS estimates that in 2012:

- only 1.2 million dwellings (5%) possessed all four of the key features that make a home fully 'visitable' (level access, flush threshold, sufficiently wide doors and circulation space, and a WC at entrance level)
- 15.7 million dwellings (69%) had at least one of these visitability features, of which 9% had three features, 21% had two and 39% had one. The remaining 5.8 million dwellings (25%) had none of the four visitability features
- older homes and terraced homes are generally more likely to be difficult to adapt, impacting on costs of any work required. Over one quarter of English homes (28%) were simply not feasible to make fully 'visitable' in 2012.

3 Literature and methodological review

3.1 Review of existing research into the cost-benefit of home adaptations

Given the comprehensive review of the evidence into the potential savings to the NHS and social care budgets from investment in housing adaptations undertaken by Heywood and Turner (2007), the BRE research project examined the summary evidence available from Heywood and Turner's research, and material published after this date.

The literature review did not seek to provide a critical analysis of the methodologies underpinning the research examined. Due to the limited resources available, the BRE research project was unable to undertake a comprehensive search and review of all the potential literature, including the cost-benefits of telecare services. However, we believe that we have captured the most recent and relevant studies for the purposes of this research.

The review was undertaken using a combination of words from Google/Google Scholar together with an examination of reference lists from the relevant literature. Virtually all the papers reviewed are related to UK studies. Details of the principal research papers studied are provided in the Appendix. This includes the main findings of the Heywood and Turner research.

3.2 Comparison of previous conclusions

A great deal of the literature examines the cost-benefit of adaptations through case scenarios, using available local and national data on the relevant costs of the adaptations and the health/social care costs saved. The London School of Economics (Snell et al, 2012) study was the first to estimate the economic benefits of an annual delivery of all DFGs in England. Not surprisingly the underpinning methodology for the studies varies a good deal; for example, a recent New Zealand study examined the benefits of home adaptations through random controlled trials.

There are commonly cited difficulties and limitations in evaluating the economic benefits of home adaptations. These include:

- incomplete evidence and the problems of isolating the benefits of adaptations alone when other sources of support, such as social care, are in place. This is especially difficult when people's needs are very complex
- home adaptations (and additional support arrangements) are geared to the needs of the individual so there is no complete uniformity in provision. Some typical or average costs may, therefore, be difficult to establish

- cross-national comparisons. These are problematic because of the differing frameworks around the delivery of home adaptations, for example whether these are wholly or partly state-funded
- time issues. The desired outcomes and costs of adaptations do not normally occur at the moment of the assessment/ decision but are generated over a period of time. Delays in installation of equipment, different lifetimes of equipment, and inflationary factors can complicate evaluations
- incorporating the costs of de-installing equipment.

Overall there is common agreement and a demonstration that home adaptations are capable of delivering significant economic benefits by reducing health and social care expenditure. Unsurprisingly, however, the picture is a complex one, not least owing to the difficulties cited above. Some particular notable findings are highlighted below.

From the evidence provided in the case studies and in the LSE study (Snell et al, 2012), the strongest supportive case for home adaptations lies around their capacity to reduce the incidence of falls in the home. This is due to the relatively high health care and social care costs of, for example dealing with a hip fracture compared to having adaptations installed. Heywood and Turner (2007) cite that health costs for dealing with hip fractures are typically five times greater than the costs of having relevant home adaptations. The New Zealand study, which trialled the effectiveness of home adaptations to prevent falls, also cited a 26% reduction in falls for the households with adaptations compared to the control group. Furthermore, for all injuries considered most relevant to the type of intervention, the rate of injuries was estimated to have fallen by 39%. While the use of randomised trials raises ethical debate, these figures translate into vast health and social care savings.

Other notable findings relating to the evaluation of DFGs:

- a stairlift and level-access shower (£6,000, the average cost of a DFG) will last at least 5 years. This expenditure would be enough to purchase the average home-care package (6.5 hours per week) for just 1 year and 3 months
- a timely DFG service could result in a delayed entry into residential care by up to 4 years. In such cases, each DFG would save around £72,000
- nationally modelling into the benefits of DFGs installed per annum (45,000 cases), estimates that the total costs of adaptations, £270 million, results in benefits of some £156 million over the lifetime of the equipment.

Although the great bulk of the literature demonstrates the positive economic cost (and social) benefits of home adaptations, both the Heywood and Turner literature review and other findings conclude that some home adaptations do not deliver economic savings. This is mainly because they cannot always remove the need for social care at home or hospital care where people are especially physically or mentally frail.

They may, however, delay a move into residential care and may lower the risks of falls in the home. Furthermore, the review of research undertaken on telecare services indicates that there is currently no overriding clear evidence that these forms of home support produce significant cost savings for the state. This is partly due to lack of available evidence but is likely to be partly related to the higher care needs of people that benefit from these services.

There is also mixed evidence on the benefits of installing additional lighting measures to prevent falls in the home. The economic benefits appear modest although this should not diminish the importance of the personal benefits, such as greater confidence, that these improvements deliver.

It is not surprising, given the complexity and difficulty of the task, that there is very little literature on quantifying the wider societal benefits of home adaptations, although the benefits to recipients and their families are well documented. These additional benefits are discussed later in this report.

4 Home interventions to reduce the risks of HHSRS hazards

This section uses some case studies provided by the Foundations Independent Living Trust (FILT) and Aster Living to demonstrate how home interventions can mitigate the risks of harm from HHSRS hazards.

4.1 Case studies: Risk of harm and HHSRS hazards

Case study 1: Reducing the risk of falls associated with stairs

Mrs D lives in a dormer-bungalow with her husband. She had problems sleeping due to her health problems so she slept on the settee. Due to her mobility problems, Mrs D struggled on steep stairs which lacked a handrail. She also had difficulties getting in and out of her bath. Her home also had an old boiler that did not work properly. It was very cold upstairs as there was no heating. The window frames were rotten.

Mrs D heard about the HIA through Age UK. Following an HIA home assessment, some home interventions were paid for by FILT and others by local authority grants. A banister rail was installed, as were a wet room and replacement windows and boiler. As a result Mrs D's general well-being has improved. She is less reliant on her husband, and feels safer, warmer and happier. She believes the banister rail helps to prevent falls and feels more confident and in control.

This case study will be revisited in Section 6.2 (Case study 3) to demonstrate the benefits to the NHS of the HIA assistance.

Case study 2: Reducing the risk of excess cold

Mrs G is an elderly woman living on her own in a house built in 1904. She contacted the HIA to have a new door fitted, but when a handyperson came to fit the new door, he suggested installing a gas fire. A caseworker then visited Mrs G to do a home assessment and asked how she coped without a fire, what her bills were like, if she was comfortable when sitting in the living room and how warm she felt. Mrs G told them, 'I have to sit with a blanket around me if I sit in the living room'.

FILT contributed £400 towards the cost of the gas fire and the client contributed £100. Mrs G also benefited from loft insulation and window replacement using other funding. The work was done during cold weather and the house is warmer. Mrs G said 'Sometimes I used to be suffering from chest colds… and sometimes when I wanted to sit and watch television a bit later I couldn't do it because I was very cold and now I could sit comfortable for a bit longer…'.

Mrs G also felt that the intervention had eased minor illnesses and symptoms of conditions, 'when it's chilly and the house is cold I seem to have a cough or a cold or arthritis in the joints,

so that extra warmth is very comforting and soothing to the pain' and may have helped to prevent other problems. 'It has done away with my general problems (diabetes, eye problems) … I haven't had to visit the doctor or the hospital for any other problems'.

The intervention also resulted in visitors staying longer in her home, 'I've had people come in (previously) and I've had to give them a blanket over their knees even when the central heating is on'. Her friends, sister and children did not visit for long because of the cold, now 'they are happy to come and to sit for a while and enjoy'.

Case study 3: Reducing the risk of excess cold

Mr C (74), who lives with his wife (64), has several medical problems including COPD and rheumatoid arthritis. When Aster Living first visited him he was using oxygen 24 hours a day, visiting his GP/hospital regularly and could not walk far within the house. The couple enquired about getting a stairlift installed (Figure 4) and Aster Living's team helped them apply for a DFG. When the caseworker visited their home to organise its installation, it turned out they were also in need of further help.

Figure 4: Mr C's stairlift. Image © Aster Living

Mr and Mrs C did not have any form of working heating as their boiler was broken (they could not afford to mend it) and their only heating, a gas fire was malfunctioning, with the potential for carbon monoxide poisoning.

Aster Living organised for temporary oil-filled radiators to be delivered so they had some form of heat while a permanent solution for them was found. Aster Living also applied for funding from charity Gas Safe to partly pay for a brand new gas fire. The remainder of the money for the fire came from Aster Living through the company's hardship fund.

Next, Charis, a company which runs financial relief schemes for major organisations such as British Gas, were called upon to replace the boiler.

There has been some improvement in Mr C's health and he now only needs the oxygen occasionally during the day.

This case study will be revisited in Section 6.2 (Case study 4) to demonstrate the cost-benefit to the NHS of Aster Living's remedial work.

5 A new cost-benefit model using the cost of poor housing approach

The BRE research project applied the methodology first developed through BRE's research published in *The real cost of poor housing* (Roys et al, 2010), in order to determine the health cost-benefits of some common home interventions that reduce the risks of harm in the home. This section outlines the elements of the modelling and how they were combined.

5.1 The Housing Health and Safety Rating System (HHSRS)

The HHSRS underpins the research methodology and is measured nationally through the EHS. The HHSRS is a risk assessment tool identifying defects in dwellings and evaluating the potential effect of any defects on the health and safety of occupants, visitors, neighbours and passers-by, particularly vulnerable people. Since its inception in 2006, the HHSRS has become the minimum standard of housing in England. The HHSRS provides a means of rating the seriousness of any hazard, so that it is possible to differentiate between minor hazards and those where there is an imminent threat of major harm or even death. Altogether 29 hazards are included as shown in Table 1, 26 of which are assessed or modelled through the EHS.

The HHSRS produces scores for dwellings based on the statistical risk of the health and safety hazard leading to harm. The scoring procedure uses a formula to generate a numerical hazard score for each of the hazards identified at the property. The higher the score, the greater the risk associated with that hazard. Potential hazards are assessed in relation to the most vulnerable class of person who might typically occupy or visit the dwelling. For example for falls on stairs and falls on the level, the vulnerable group is older people (aged 60 or over).

Table 1: The 29 HHSRS hazards

Physiological requirements	Protection against infection
Damp and mould growth etc.	Domestic hygiene, pests and refuse
Excessive cold	Food safety
Excessive heat	Personal hygiene, sanitation and drainage
Asbestos etc.	Water supply
Biocides	
CO and fuel combustion productions	**Protection against accidents**
Lead	Falls associated with baths etc.
Radiation	Falling on level surfaces
Un-combusted fuel gas	Falling on stairs etc.
Volatile organic compounds	Falling between levels
	Electrical hazards
Psychological requirements	Fire
Crowding and space	Flames, hot surfaces etc.
Entry by intruders	Collision and entrapment
Lighting	Explosions
Noise	Position and operability of amenities etc.
	Structural collapse and falling elements

The hazard score formula requires the surveyor to make two judgements:

- The likelihood of the occurrence which could result in harm to a vulnerable person over the following 12 months. The likelihood is to be given as a ratio, for example 1 in 100, 1 in 500.
- The likely health outcomes or harms which would result from the occurrence. From any occurrence there may be a most likely outcome, and other possible ones which may be more or less severe. For example a fall from the top of a flight of stairs could result in a 60% chance of a severe concussion or fracture, but there may also be a 30% chance of a more serious injury and a 10% chance of something less serious. The four classes of harms and associated weightings are listed in Table 2.

From the judgements made by the surveyor, a hazard score can be generated for each hazard as illustrated in Table 3, using the example of falling between levels.

Using this method, hazard scores can range from 1 (very safe) to over 5,000 (very dangerous). A score of 1,000 or more is considered to be a Category 1 hazard, a serious hazard. In the previous BRE research published in *The real cost of poor housing* (Roys et al, 2010), poor housing is defined as a home that has 1 or more Category 1 HHSRS hazards.

5.2 Defining and quantifying national eligibility for home adaptations using the EHS

The EHS can identify both the homes with serious HHSRS hazards (Table 1) and the types of households who live in these homes. Using EHS data, the national group of households considered most likely to benefit from home interventions, and therefore most suited to the modelling, were households

which contained someone with a long-term sickness or disability that limits their activity. Information on these households is collected through the EHS interview survey and the assessment of whether any long-term sickness and disability exists is undertaken by respondents.

In 2012 the EHS estimated that some 6.4 million households contained someone with a long-term sickness or disability. Of these, around 854,000 (13%) lived in a home with at least one of the Category 1 hazards assessed by the EHS.

The 2011 EHS also estimates that:

- around 726,000 households contained at least one person who used a wheelchair at least some of the time. Three quarters (75%) of these people lived in older households, where the oldest person was aged 60 or over
- some 15% of those households with a long-term sickness felt that their current home was not suitable for their needs
- those adaptations most commonly needed, as assessed by the household, were: grab rails inside the dwelling; a bath/shower seat or other aids to use a bath/shower; a shower to replace the bath; and a special toilet seat.

5.3 What the research included

Many of the 26 HHSRS hazards collected by the EHS cannot be remedied through more common home adaptations, so these hazards were excluded from the model. However, these excluded hazards were far less common in homes; the most common hazards are falls associated with stairs and excess cold. In total, 11 HHSRS relevant to common home adaptations were included in the modelling, grouped into four key areas: risks associated with falls, kitchens, bathrooms and heating. The links between these four groups, and some of the most common types of home interventions, are provided in Table 4.

Table 2: Classes of harms and weightings used in the HHSRS

Class	Examples	Weightings
Class 1	Death, permanent paralysis below the neck, malignant lung tumour, regular severe pneumonia, permanent loss of consciousness, 80% burn injuries	10,000
Class 2	Chronic confusion, mild strokes, regular severe fever, loss of hand or foot, serious fractures, very serious burns, loss of consciousness for days	1,000
Class 3	Chronic severe stress, mild heart attack, regular and persistent dermatitis, malignant but treatable skin cancer, loss of a finger, fractured skull, severe concussion, serious puncture wounds to head or body, severe burns to hands, serious strain or sprain injuries, regular and severe migraine	300
Class 4	Occasional severe discomfort, chronic or regular skin irritation, benign tumours, occasional mild pneumonia, a broken finger, sprained hip, slight concussion, moderate cuts to face or body, severe bruising to body, 10% burns, regular serious coughs and colds	10

Table 3: Example hazard score for falls between levels

Class	Weighting		Likelihood (1 in)		Spread of harm		Hazard score
Class 1	10,000	÷	100	×	0	=	0
Class 2	1,000	÷	100	×	30	=	300
Class 3	300	÷	100	×	60	=	180
Class 4	10	÷	100	×	10	=	1
All classes							481

Table 4: Links between key HHSRS hazards groups and common types of home adaptations

Type of adaptation	Falls	Kitchens	Bathrooms	Heating
Extension of home	▓			
Redesign kitchen		▓		
Redesign bathroom			▓	
Graduated floor shower			▓	
Stairlift	▓			
External ramp	▓			
New bath/shower room			▓	
Shower replacing bath			▓	
Wheelchair accessible parking	▓			
Adjustable bed or related aid	▓			
Hoist	▓		▓	
Wide doorways	▓			
Additional relocated toilet	▓		▓	
Low level bath			▓	
Relocate bath/shower			▓	
Additional heating				▓
Shower over bath			▓	
Wide paths	▓			
Entry phone				
Other external adaptation	▓			
Other modification of kitchen		▓		
Individual alarm system	▓			
External rail to steps	▓			
Internal ramp	▓			
Bath/shower seat			▓	
Visual/hearing impairment related				
Wide gateway	▓			
Electrical modifications				
Grab rail or other rail	▓			
Toilet seat			▓	

EHS 2012 data estimates that there were around 875,000 Category 1 hazards for the relevant HHSRS hazards used in the modelling, in the homes of households containing someone with a long-term sickness or disability (Figure 5). In addition, the survey estimates that there were around 2.2 million high-scoring Category 2 hazards (worse than average) in homes occupied by such households. **In total, therefore, around 3 million homes had significantly higher than average risks.** These findings were used for the national model.

5.4 Quantifying the cost to society of poor housing

5.4.1 What costs should be included?

This is a key question as some types of cost can be estimated or modelled more reliably than others. One of the most comprehensive reviews of poor housing (Ambrose, 2001) provides a matrix of costs, categorising them in terms of their measurability – costs that can be quantified (H); costs that could be quantified given better data (M); and costs that exist but are probably non-quantifiable (NQ) (Table 5).

Following a review of data sources and an attempt to cost up all of these factors and link them directly to hazards in the home, it was decided that we should focus on the NHS treatment costs alone (highlighted in Table 5). This is because:

- it is a transparent method of selecting a typical outcome for each level of harm of each hazard
- robust data is available to estimate the medical and care costs for these NHS treatment costs, shown in Table 5
- we cannot be accused of overstating the case by making heroic assumptions.

Having determined this methodology, our next step was to provide descriptions of the NHS treatments for the different outcomes for different hazards. These could then be calculated, using NHS data (Table 6).

The representative treatment costs have all increased from the published 2004 values (Roys et al, 2010). If we had assumed a 3% annual growth on the 2004 prices, then the increase in representative costs would have given values of £61,494, £24,597, £1,845 and £123, respectively. The representative cost estimates for Class 3 and 4 harms are very similar to those provided through a 3% annual increase, whereas those in Class 1 and 2 have increased more than expected. This is mainly due to the increased cost applied to bed days for spinal injury; in the 2004 values, a standard bed day plus treatment costs were applied. These revised representative costs have been used for the rest of the modelling.

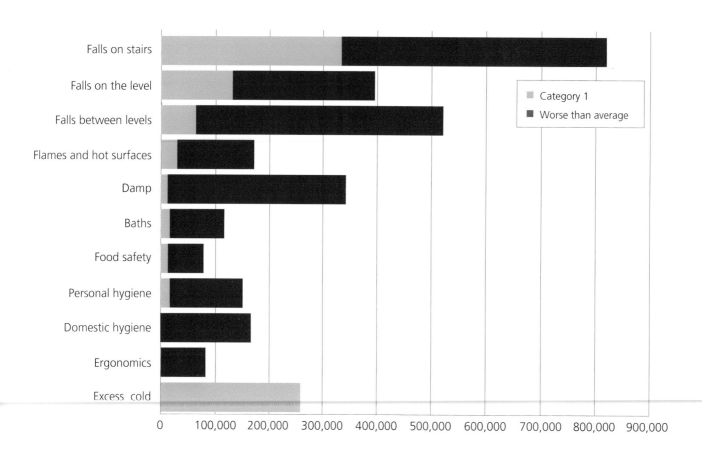

Figure 5: Frequency of worse than average and Category 1 hazards in households containing someone with a long-term sickness or disability

Table 5: The costs to society of poor housing

Resident's costs	Area
Annual loss of asset value if owned (H)	Annual loss of asset value if rented (H)
Poor physical health (H)	Higher health service treatment cost (H)
Poor mental health (M)	Higher care service treatment cost (M)
Social isolation (NQ)	Higher building heating costs (H)
Higher insurance premiums (H)	Uninsured external losses (M)
Uninsured content losses (M)	Extra school costs/homework classes (H)
Underachievement at school (NQ)	Loss of talents to society (NQ)
Loss of future earnings (M)	High policing cost (H)
Personal insecurity (NQ)	High emergency service costs (H)
More accidents (M)	High environmental health costs (H)
Poor hygienic conditions (NQ)	Disruption to service providers (M)
Costs of moving (M)	Special health care responses (H)
Adopting self-harming habits (M)	Government and EU programmes (H)

H: Costs that can be quantified.

M: Costs that could be quantified given better data.

NQ: Costs that exist but are probably non-quantifiable.

Table 6: Typical outcomes and first year treatment cost for selected HHSRS hazards

Hazard	Class 1	Class 2	Class 3	Class 4
Damp and mould growth	–	£2,034	£1,027	£242
Excess cold	£19,851	£22,295†	£519	£84
Radon	£13,247	£13,247†	–	–
Falls on the level	£92,490*	£39,906†	£1,545	£115
Falls on stairs	£92,490*	£39,906†	£1,545	£115
Falls between levels	£92,490*	£6,464†	£2,476	£115
Fire	£14,662†	£7,435†	£1,879	£123
Flames and hot surfaces	–	£7,378	£1,822	£123
Collision and entrapment	–	£5,152	£1,698	£115
Representative cost	**£90,000**	**£30,000**	**£1,800**	**£120**
3% annual increase from 2004 prices	**£61,494**	**£24,597**	**£1,845**	**£123**

*   Costs after the first year will occur.

†   Costs after the first year are likely to occur, as a consequence of the initial illness/incident.

5.5 Estimating the cost to the NHS of HHSRS hazards

Using these representative costs, it is possible to calculate the estimated cost to the NHS of not mitigating each of the hazards. Since the mitigation only reduces the risk to average for the age of the property, the cost calculation also considers the costs to the NHS in its existing condition minus the costs to the NHS of the average condition. This calculation can be shown as:

$$\sum_{Class\ 1}^{Class\ 4} \frac{Cost\ weighting_{(class)}\ *\ Spread\ of\ harm_{(class)}}{Likelihood_{(before)}\ -\ Likelihood_{(after)}}$$

Applying the likelihood values and harm distributions in Tables 7, 8 and 9 with the representative costs from Table 6, the total cost to the NHS of not mitigating the harms can be calculated using this formula.

In total, the cost to the NHS of not mitigating these hazards in homes of long-term sick and disabled people is calculated to be nearly £414 million (Table 10).

The majority of these costs are associated with excess cold, falls on stairs and falls on the level.

Table 7: Distribution of likelihood and harms, Category 1 hazards

Type of hazard	Number	Average likelihood	Class 1	Class 2	Class 3	Class 4
Falls on stairs	333,989	33	4.4%	16.6%	27.8%	51.2%
Falls on the level	131,172	19	1.0%	16.1%	38.1%	44.9%
Falls between levels	63,944	65	13.1%	8.0%	18.0%	60.9%
Flames and hot surfaces	29,077	6	0.1%	2.4%	23.2%	74.4%
Damp	12,187	2	0.0%	1.0%	10.0%	89.0%
Baths	15,669	18	2.2%	4.6%	10.0%	83.2%
Food safety	12,698	9	0.0%	2.0%	22.0%	76.0%
Personal hygiene	15,882	9	0.0%	2.0%	22.0%	76.0%
Domestic hygiene	551	1	0.0%	0.1%	1.0%	98.9%
Ergonomics	454	8	0.0%	1.7%	16.9%	81.4%
Excess cold	259,589	49	34.0%	6.0%	18.0%	42.0%

The data in the cells shaded in blue are based on HHSRS guidance data, with likelihoods leading to a HHSRS score of 1000. Excess cold likelihood is calculated from the average SAP value for the sample of homes.

Table 8: Distribution of likelihood and harms, not Category 1 but worse than average hazards

Type of hazard	Number	Average likelihood	Class 1	Class 2	Class 3	Class 4
Falls on stairs	485,145	91	2.4%	12.2%	24.2%	61.1%
Falls on the level	263,836	45	0.2%	11.7%	33.8%	54.3%
Falls between levels	456,620	322	3.0%	5.5%	16.0%	75.5%
Flames and hot surfaces	141,913	48	0.1%	1.9%	24.6%	73.3%
Damp	330,474	117	0.0%	1.0%	10.0%	89.0%
Baths	100,782	53	1.9%	3.6%	10.3%	84.2%
Food safety	66,333	19	0.0%	2.0%	22.0%	76.0%
Personal hygiene	134,849	19	0.0%	2.0%	22.0%	76.0%
Domestic hygiene	167,942	3	0.0%	0.1%	1.0%	98.9%
Ergonomics	81,064	15	0.0%	1.7%	16.9%	81.4%
Excess cold	–	744	34.0%	6.0%	18.0%	42.0%

The data in the cells shaded in blue are based on HHSRS guidance data, with likelihoods leading to a HHSRS score of 500.

Table 9: Distribution of likelihood and harms, average values

Type of hazard	Average likelihood	Class 1	Class 2	Class 3	Class 4
Falls on stairs	245	1.9%	6.7%	21.7%	69.7%
Falls on the level	135	0.2%	13.8%	27.3%	58.7%
Falls between levels	1693	0.2%	1.8%	9.9%	88.1%
Flames and hot surfaces	182	0.0%	1.3%	17.8%	80.9%
Damp	464	0.0%	1.0%	10.0%	89.0%
Baths	4026	1.9%	3.6%	10.3%	84.2%
Food safety	4960	0.0%	2.0%	22.0%	76.0%
Personal hygiene	7750	0.0%	2.0%	22.0%	76.0%
Domestic hygiene	5585	0.0%	0.1%	1.0%	98.9%
Ergonomics	12925	0.0%	1.7%	16.9%	81.4%
Excess cold	2152	34.0%	6.0%	18.0%	42.0%

The data in the cells shaded in blue are based on HHSRS guidance data.

Table 10: Total cost to the NHS associated with not mitigating potential harms

Type of hazard	Total benefit potential (2011 cost)	Category 1 hazards				Significantly worse than average (but not Category 1)			
Falls on stairs	£115,147,096	£37,093,972	£47,222,248	£4,479,219	£500,485	£8,419,279	£15,642,107	£1,562,460	£227,326
Falls on the level	£56,105,215	£5,949,867	£29,239,957	£4,242,123	£302,544	£969,466	£12,531,038	£2,623,940	£246,280
Falls between levels	£20,655,296	£11,497,783	£2,341,024	£310,006	£67,554	£3,798,537	£2,179,816	£360,672	£99,905
Flames and hot surfaces	£9,270,359	£494,309	£3,378,470	£1,969,082	£416,923	£344,728	£1,415,710	£1,065,653	£185,484
Damp	£4,797,718	£-	£1,820,170	£1,092,102	£647,981	£-	£632,651	£379,590	£225,224
Baths	£8,929,491	£1,716,935	£1,197,087	£155,968	£86,517	£3,208,840	£2,026,636	£347,906	£189,603
Food safety	£5,312,239	£-	£844,997	£557,698	£128,440	£-	£2,086,702	£1,377,223	£317,179
Personal hygiene	£9,613,602	£-	£1,057,570	£697,996	£160,751	£-	£4,247,950	£2,803,647	£645,688
Domestic hygiene	£9,417,670	£-	£16,527	£9,916	£65,381	£-	£1,678,518	£1,007,111	£6,640,217
Ergonomics	£4,974,041	£-	£28,925	£17,253	£5,540	£-	£2,752,977	£1,642,070	£527,276
Excess cold	£169,676,603	£158,419,500	£9,318,794	£1,677,383	£260,926	£-	£-	£-	£-
Total	**£413,899,331**	**£215,172,365**	**£96,465,771**	**£15,208,747**	**£2,643,042**	**£16,740,849**	**£45,194,105**	**£13,170,271**	**£9,304,181**

5.6 Estimating the costs to make homes safer

The EHS also collects details on the types of work needed to mitigate HHSRS hazards to an acceptable level; this level is usually the average for the age and type of dwelling. These works are costed up using standard prices. For hazards that are modelled (apart from excess cold), we used costs for a 'typical' package of work. For excess cold, we used BRE's Energy Upgrade Model to determine the most cost-effective works needed to raise the SAP rating to the national average, and then used standard prices for these. The average costs to mitigate the hazards used in the modelling are provided in Figure 6.

The average cost of remedial work for all these hazards was around £2,130.

5.7 Estimating the cost of adaptations as a result of an accident from HHSRS

While the total DFG budget per annum and the average cost per DFG is known, it is difficult from this information alone to allocate the appropriate adaptation cost to the potential hazard that led to the referral for an adaptation. However, estimates exist for the number of adaptations and the average adaptation costs for each adaptation type, based on data collected by DCLG. The accuracy and quality of this data is unknown, which implies that better recording of such data could improve any cost modelling in the future. In the absence of better data this information has been used to estimate the potential costs of adaptations. The overall cost is estimated to be £15 billion, which is nearly 10 times the DFG budget. Given the list of

adaptations, and the average cost of just under £2,700, it is likely that this list covers adaptations provided by handyperson services, other social service methods and payments by individuals.

The costs associated with each of the individual adaptations have been distributed according to the distribution of hazards shown in Table 4. The percentages shown in each column of Table 11 show the distribution of costs across the types of adaptations. For example the adaptations related to kitchens are distributed such that 62% is related to redesigning kitchens, and 38% is related to other kitchen modifications.

The average costs of adaptations that can be applied in the model are calculated to be £2,779 for falls-related adaptation, £5,622 for kitchen-related adaptations, £2,584 for bathroom-related adaptations and £1,775 for heating adaptations. It is also clear from Table 11 that the majority of adaptations are distributed to falls or bathrooms. Kitchen and heating adaptations, between them, only account for 5% of all of the adaptation.

These average costs can be applied to a proportion of the dwellings where a referral has been made following harm caused by one of the HHSRS hazards. It is assumed that referrals come from many different sources, and not all the referrals will follow such an event.

Considering the likelihoods and number of households with someone at risk, the total number of potential referrals from hazardous events is 131,101. This equates to just 2.3% of all the adaptation referrals considered by the model. It has been assumed for this model that every hazardous event leading to harm is referred for adaptations. This is likely to be an overestimate, but even at this rate it is less than 1% of the total adaptations cost.

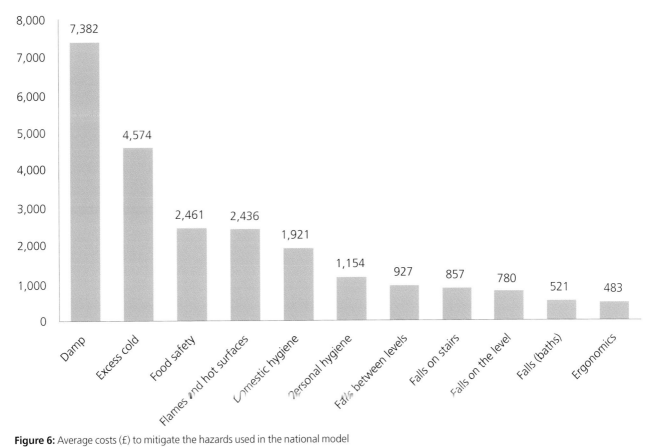

Figure 6: Average costs (£) to mitigate the hazards used in the national model

Table 11: Distribution of adaptation costs by type of hazard

Type of adaptation	Number	Percentage of total	Fall	Kitchen	Bathroom	Heating	Typical cost	Total cost
Extension of home	83,000	1.47%	2.81%				£27,500	£2,282,500,000
Redesign kitchen	114,000	2.01%		61.62%			£8,500	£969,000,000
Redesign bathroom	185,000	3.27%			7.62%		£7,550	£1,396,750,000
Graduated floor shower	214,000	3.78%			8.81%		£5,750	£1,230,500,000
Stairlift	276,000	4.87%	9.36%				£5,400	£1,490,400,000
External ramp	265,000	4.68%	8.98%				£4,500	£1,192,500,000
New bath/shower room	154,000	2.72%			6.34%		£4,500	£693,000,000
Shower replacing bath	332,000	5.86%			13.67%		£4,250	£1,411,000,000
Wheelchair accessible parking	165,000	2.91%	5.59%				£4,250	£701,250,000
Adjustable bed or related aid	250,000	4.41%	8.47%				£3,250	£812,500,000
Hoist	65,000	1.15%	2.20%		2.68%		£3,150	£204,750,000
Wide doorways	125,000	2.21%	4.24%				£2,625	£328,125,000
Additional relocated toilet	156,000	2.75%	5.29%		6.43%		£2,500	£390,000,000
Low level bath	111,000	1.96%			4.57%		£2,000	£222,000,000
Relocate bath/shower	67,000	1.18%			2.76%		£1,900	£127,300,000
Additional heating	86,000	1.52%				100.00%	£1,775	£152,650,000
Shower over bath	188,000	3.32%			7.74%		£1,700	£319,600,000
Wide paths	145,000	2.56%	4.92%				£1,275	£184,875,000
Entry phone	114,000	2.01%					£1,250	£142,500,000
Other external adaptation	93,000	1.64%	3.15%				£1,000	£93,000,000
Other modification of kitchen	71,000	1.25%		38.38%			£1,000	£71,000,000
Individual alarm system	143,000	2.53%	4.85%				£850	£121,550,000
External rail to steps	319,000	5.63%	10.81%				£775	£247,225,000
Internal ramp	43,000	0.76%	1.46%				£505	£21,715,000
Bath/shower seat	546,000	9.64%			22.49%		£483	£263,718,000
Visual/hearing impairment related	63,000	1.11%					£475	£29,925,000
Wide gateway	84,000	1.48%	2.85%				£275	£23,100,000
Electrical modifications	58,000	1.02%					£275	£15,950,000
Grab rail or other rail	738,000	13.03%	25.02%				£140	£103,320,000
Toilet seat	410,000	7.24%			16.89%		£40	£16,400,000
		100%	100%	100%	100%	100%		£15,258,103,000
Total	**5,663,000**		**2,950,000**	**185,000**	**2,428,000**	**86,000**		
Average cost			**£2,779**	**£5,622**	**£2,584**	**£1,775**	**£2,694.35**	

6 Estimating the reduction in cost to the NHS of home interventions

6.1 Applying the cost of poor housing model to calculate the health cost-benefit to the NHS

The cost-benefit analysis for home interventions/adaptations assumes that the cost to the NHS for treating harms and the cost of work can be mitigated if improvements are made to homes where there is a risk of harm in the next 12 months. These costs and savings are summarised in Table 12.

In total, mitigating the risk associated with these hazards in homes occupied by someone at risk of harm would cost £6.4 billion. This is a huge cost, but should benefit people in all 3 million houses to which it is applied. The total benefit to the NHS by making these improvements is nearly £414 million per annum. The ratio between the cost to mitigate these hazards,

and the benefit to the NHS, enables a simple calculation of economic payback for the intervention. It is important to recognise that the payback period relates to the NHS and not to the individual in receipt of the home interventions. Indeed the mitigation work suggested is intended to impact many homes where the recipients would never have received any NHS treatment at all, even without the intervention.

However, in this model we can also offset the benefits gained by not having to make adaptations to these homes. Payback is therefore calculated as total cost to mitigate the hazard minus the benefits from adaptations all divided by total benefit to the NHS. On average the payback period is 15 years. The best paybacks come from mitigating falls on the level (5.2 years), falls on stairs (5.9 years), falls in baths (6.5 years) and excess cold (6.9 years). Benefits from mitigating damp conditions are not cost-effective.

Table 12: Total cost-benefit from mitigating hazards in vulnerable homes

Type of hazard	Number of cases	Total cost to mitigate	Total benefit potential (2011 cost)	Benefit to adaptation budget	Total benefit	Payback (years)
Falls on stairs	819,134	£701,997,838	£115,147,096	£27,815,080	£142,962,176	5.9
Falls on the level	395,008	£308,106,240	£56,105,215	£19,151,301	£75,256,517	5.2
Falls between levels	520,564	£482,562,828	£20,655,296	£2,720,910	£23,376,206	23.2
Flames and hot surfaces	170,990	£416,531,640	£9,270,359	£27,246,284	£36,516,643	42.0
Damp	342,661	£2,529,523,502	£4,797,718	£10,818,784	£15,616,502	525.0
Baths	116,451	£60,670,971	£8,929,491	£2,251,656	£11,181,147	6.5
Food safety	79,031	£194,495,291	£5,312,239	£7,934,975	£13,247,214	35.1
Personal hygiene	150,731	£173,943,574	£9,613,602	£4,567,771	£14,181,373	17.6
Domestic hygiene	168,493	£323,675,053	£9,417,670	£3,153,494	£12,571,164	34.0
Ergonomics	81,518	£39,373,194	£4,974,041	£324,431	£5,298,472	7.9
Excess cold	259,589	£1,187,360,086	£169,676,603	£9,403,479	£179,080,082	6.9
Total	3,104,170	£6,418,240,217	£413,899,331	£115,388,164	£529,287,495	15.2

6.2 Case studies: The health cost-benefit of individual interventions

Case study 1: Health benefits of prevention of falls to stairs

This 1920s mid-terrace home shown in Figure 7a is occupied by a widow, Mrs M, in her mid-70s with growing mobility problems; she will need both knee and hip replacements in the next 12 to 18 months. The main staircase is fitted with a handrail to one side, which is difficult to grasp (Figure 7b). At the top of the main stairs there are a further two steps to the right, and to the rear, there is a step down to the corridor to the bathroom and an additional bedroom (Figure 7c). All stair treads are narrow but not exceptionally so for the age of the property. The busy carpet pattern makes it difficult to determine the edge of the steps (Figure 7b and 7c). The property also has a cellar accessed by a short stone staircase with no handrail (Figure 7d). The cellar is used on a regular basis to access the electricity and gas meters and is also regularly used for storage (Figure 7d). The back door has two steps to the rear yard. These are of uneven height and the paving stone serving as one of the treads is loose.

The property has a significantly higher average risk of a fall from the stairs due to the lack of handrails to the cellar stairs; any fall from these stairs would result in Mrs M landing on a hard and

Figure 7: 1920s terraced homes with a Category 1 hazard associated with falls on stairs

unforgiving surface. Furthermore, the loose and wobbly steps to the back garden increase the chance of a fall. Even the main staircase presents a higher than average risk, with the poor handrail and visually complicated carpet pattern. It is evident that Mrs M would benefit from additional support on all stairs due to the growing weakness in her knees and legs.

The costs of installing two handrails and undertaking repair work to the rear steps would be in the region of £350 to £450. The addition of graspable handrails to the main stairs at a cost of £200 would also reduce the risk of a fall. Changing the carpet on the stair should make it easier to see the edge of steps, further reducing the risk, but the same could be achieved by painting white stripes on the step nosings.

It is estimated that the likelihood of harm in the next year for these stairs is 1 in 10, with a higher chance of a severe outcome given the tiled surface on the cellar stairs. Using the national model, the mitigations suggested above would result in typical health cost savings of £932. If the prevention of a fall also removed the need for future home adaptations, then this benefit could increase by £315 to about £1,247. Consequently the remedial work would likely pay for itself within the first year. It is likely that remedial action preventing a fall would reduce the risk of early provision of home-care services that Mrs A would likely need following a very serious fall, for example help with cooking and personal care costing around £85 per day, as well as weekly follow-up visits from her GP's surgery, costing £25 per week.

Case study 2: Health benefits of prevention of falls to stairs

This three-bedroom house is owned by a couple both aged over 80 years with growing mobility problems; they pay a neighbour to maintain their lovely garden and pay for help with cleaning their home. They both have arthritis which has become more painful recently and has made gripping items such as cooking utensils very difficult; consequently they pay for their meals to be delivered during the week and weekend hot meals are provided by their family. The home has two internal trip steps on the ground floor from the kitchen to the rear garden room (Figure 8a) and further external steps outside the rear door (Figure 8b and 8c). These steps are in need of some repair as the uneven nature increases the risks of falls. For some of the external steps the rear doors opens over the flight, making navigation and the addition of handrails difficult. The last few steps are also quite narrow, increasing the risk of falling off the side of the steps (Figure 8c). Furthermore, the steps are especially slippery in wet weather. Together the nature of these stairs and trip steps at the home mean that the risk associated with a fall is significantly worse than average, estimated to have a likelihood of 1 in 6, with a higher chance of a severe outcome given the potential collision with stone slabs and the isolated nature of the stairs.

The cost to mitigate this risk through repair to the steps to even out the variance in step dimensions and the installation of grab and handrails is estimated to be £2,550. The national modelling estimates that the cost-benefits to the NHS in the first year are £1,565. If the prevention of a fall also removed the need for future home adaptations, then this benefit could increase by £525 to about £2,090. This means that the remedial work would pay for itself in around 15 months.

Any fall for either occupant would likely lead to greater fragility and expedite the need for help with personal care, as both husband and wife lack the ability to care for the other in this respect. This could result in a care package of around £85 per day, over £31,000 per annum.

Figure 8: Stairs and trip steps increase the risk of a harmful fall

Case study 3: Health benefits of prevention of falls to stairs and excess cold

Returning to our example of home interventions in Section 4.1 (Case study 1) of this report, due to her health problems, Mrs D had the following home interventions and adaptations: a handrail installed on her stairs, a wet room, replacement windows and a new boiler. The total cost of these works is estimated at £12,200.

These adaptations, as well as making it easier for Mrs D to care for herself in her own home, should also reduce the risk of harm from falls on stairs, falls in bathrooms, excess cold and personal hygiene. Assuming that in each case the harm was worse than average, then the potential savings to the NHS in the first year would be in the region of £820. This would mean that the payback period for this work is 14.9 years; however, these savings do not take into account any additional home-care savings that may have arisen as a result of Mrs D having a serious fall.

Case study 4: Health benefits of prevention of excess cold

Returning to our example of home interventions in Section 4.1 (Case study 3) of this report, Mr C (74) and Mrs C (64) enquired about getting a stairlift, but subsequent investigations found that their only heating, a gas fire, was malfunctioning with a risk of carbon monoxide poisoning.

The cost of the works to improve their home by installing a new condensing boiler and a gas fire is estimated to be £1,800. These improvement works reduced the risk of the couple suffering harm from excess cold, for which the annual savings to the NHS alone are estimated to be £650, giving a payback period of 2.8 years. The improvements also reduced the risk of carbon monoxide poisoning, which would provide a further annual benefit of £97, as estimated from the national model, reducing the payback to 2.4 years.

7 Additional costs to society of not adapting homes for the disabled and vulnerable

7.1 Review of 'unquantifiable' costs

The ongoing cost to the NHS of not adapting homes when required only represents part of the total costs to society of not investing in home adaptations. An unsafe and unsuitable home has knock-on effects on other areas of government spending, particularly in relation to home-care costs, as well as implications for the occupants themselves. For young disabled people, a serious home accident may result in loss of potential earnings and loss of potential revenue for the state. These and other examples are detailed in the Table 13 shown on page 28.

One of the most important benefits of home adaptations is the positive impact on the well-being, independence, privacy and dignity of the people who require the adaptations (Figure 9). Illustrative quotations are provided from Heywood's research into the effectiveness and value of housing adaptations (Heywood, 2001) and from the Papworth Trust (Papworth Trust, 2012) together with feedback from Habinteg Housing Association tenants.

> 'Believes she wouldn't be alive without the adaptations – for psychological rather than physical reasons. Hated having to ask for help and her life was existing, not living.'

> 'He wouldn't be at home with his family without it. We wouldn't be as happy as we are without it. It has definitely transformed both our lives.'

> 'The change has made a huge difference. My husband had to carry me upstairs every day, in the end he was totally exhausted. I am now able to go to the bathroom by myself.'

> 'It has improved the quality of my wife's life. For the first 15 months in this property, she had to go weekly to a Leonard Cheshire home to have a bath and had to use a commode in her bedroom, as her wheelchair was too big to fit in the bathroom.'

> 'It has made life a great deal easier. [Without it,] it would have been impossible for me to keep on living in our home or even to go out.'

Figure 9: Bathroom sink without pedestal (a) and bathroom hoist (b) to enable wheelchair user to access washing facilities. Prior to adaptations, the householder said, 'I couldn't have many baths because it meant organising extra assistance. I would occasionally check in to an accessible hotel to have a bath!' Images © Habinteg Housing Association

Table 13: Additional benefits to society of home adaptations

Beneficial outcomes for government/economy	Beneficial outcomes for households – financial, physical, emotional well-being and social	Other beneficial outcomes for families
Reduction in (state-funded) care home	Reduction in (self-funded) care home	Reduced risk of physical injury, such as back strain or musculoskeletal problems where the required appropriate technology does not exist to help with the task of caring
Reduction in hospitalisations and bed-blocking due to accidents/cold homes	Physical health – safety and avoidance of accidents. Reduction in stress and anxiety	Likely reduction in time off work to provide care and support. Reduction in loss of earnings
Reduction in demand for other health and adult care services, for example mental health services	Improved nutrition, for example kitchen alterations for better positioned cooking facilities and work surfaces may lead to improved nutrition by people eating fewer ready meals and takeaways	Reduced emotional stress and anxiety
Reduction in emergency services costs, for example following a fall	Greater privacy, independence, autonomy and control	Cost savings (through reduction in need to travel)
Fall in productivity losses, for example through added time off work to recuperate/provide care Reduction in losses of taxation	Improved self-confidence, peace of mind and sense of safety in the home, improved psychological well-being	–
Reduced potential additional costs to NHS, for example back strain or musculoskeletal problems among carers who lack the required appropriate technology to help them with the task of caring	Enhanced family relationships and other relationships	–
Reduction in associated welfare benefits	Reduction in social isolation, for example ramps and level access adaptations allow opportunities for involvement outside the home	–
Reduce potential demand for general housing waiting lists. If a household's accommodation is unsuitable for their needs or if the household does not feel safe, this may result in an application for or a move into scarce social housing	For children – reduction in school absenteeism and in cases of serious injury, loss of potential lifetime earnings	–

8 Conclusions

This research has demonstrated some of the cost-benefit to the NHS of undertaking preventative home interventions for households with a long-term sickness or disability, where the risk of accidents in their home are worse than the national average. Furthermore, it has been possible to demonstrate how the cost of this preventative action is offset through savings to the DFG budget, therefore providing an additional payback for the work to the state and society.

A new national model was created to quantity these benefits using the same basic methodology developed to calculate the data published in *The real cost of poor housing* (Roys et al, 2010), together with enhanced modelling improvements. A number of assumptions have been tested during the costing exercise, but our best estimate suggests that leaving long-term sick and disabled households in homes with significant hazards is costing the NHS nearly £414 million per annum in first year treatment costs. Furthermore, if we add the costs of installing a home adaptation following a harmful event, such as a fall on stairs, because remedial action has not been undertaken, the cost rises to around £529 million per annum. These savings to the DFG budget are important given the increasing pressures for adaptations as a result of, for example, our ageing society and the pressures on public expenditure. It is important to note that there are potentially additional savings resulting from such a preventative approach, including savings to social care.

The bulk of these costs to the NHS arise from a lack of remedial action to address Category 1 hazards. The largest costs occur due to the treatment of harms arising from excess cold. Although excess cold comprises 8% of all the hazards identified among the homes of long-term sick and disabled households, it comprises 34% of the £529 million cost to the NHS identified in this research. In addition, lack of remedial action to address the risk of falls, particularly those associated with stairs, incurs notably higher NHS treatment costs. Falls on stairs comprise 38% of Category 1 hazards and 22% of other significantly worse than average hazards and comprise around a quarter (24%) of the NHS costs identified.

The total cost of remedial works to mitigate the risk associated with these hazards in homes occupied by someone at risk of harm is estimated to be £6.4 billion. Although a huge cost, the expenditure should benefit people in all 3 million houses to which it is applied. Furthermore, the average cost of work per household is just £2,130. On average the payback period to the NHS to mitigate each type of harm is 15 years. The best paybacks come from mitigating falls on the level (5.2 years), falls on stairs (5.9 years), falls in baths (6.5 years) and excess cold (6.9 years).

This research has identified the need for preventative work in around 3 million households who have a long-term sickness and disability. This work would make their homes safer and warmer and so reduce the likelihood of NHS treatment and the need for a DFG adaptation required as a result of injury. However, these 3 million households represent only a proportion of households, who are likely to need adaptations to their homes.

For example, over half of the 6.4 million long-term sick and disabled households investigated for this research have been identified as living in a home where the risk of harm, as assessed through the HHSRS, does not exceed the national average. Nonetheless, a proportion of these households will still require a home adaptation due to their difficulties in maintaining independence on a daily basis because of their physical and medical circumstances. Furthermore, as previous BRE research for DCLG cited, while an estimated 1 million households require adaptations to their home there is no robust and definitive means to establish the potential demand for DFGs in the future, let alone the need for adaptations paid by households themselves or other charitable sources (BRE for DCLG, 2011).

This research therefore demonstrates that simple and relatively low cost home safety improvements, such as handrails on dangerous stairs and steps, are very cost-effective on health considerations alone. The health cost gains of mitigating cold homes have been demonstrated, but these works also deliver fuel cost savings to households as well as carbon savings. Through the use of case studies, this research has also demonstrated the importance of home interventions to the quality of life for those receiving them.

The literature review undertaken for this report highlights the difficulties in obtaining robust estimates of the economic benefits of home adaptations, but there is common agreement and evidence from other research that home adaptations are able to deliver significant economic benefits by reducing health and social care expenditure, particularly in preventing falls in the home. The new national model is unable to estimate the savings of home interventions to the social care budget, and so it is important to recognise that the savings to the NHS estimated in this research represent only a proportion of potential savings to public expenditure.

It is hoped that this research will support a more informed case for investment in housing interventions and adaptations, on the basis that it not only improves people's health and quality of life, but also makes sound economic sense; it saves public money in the longer term.

In the future it may be possible to adapt or enhance the methodology used for the report. It could be developed into a practical tool to enable local housing and health providers and to demonstrate the value of housing interventions, especially where there are perceived risks to the safety of people in their homes.

It is evident, however, that further research is required into the full economic benefits of home adaptations such as DFGs, at a national level, particularly into the potential wider costs savings to NHS/social care budgets so that these are better understood.

Appendix: Literature review

Heywood F and Turner L (2007). Better outcomes, lower costs: Implications for health and social care budgets of investment in housing adaptations, improvements and equipment – a review of the evidence. Office for Disability Issues, Department for Work and Pensions

This research reviewed the evidence, through a comprehensive review of the literature, regarding the implications for health and social care budgets of investment in housing adaptations, improvements and equipment. The authors acknowledge that the evidence is not complete and that it is not always possible to isolate single factors in determining the cost-benefits of adaptations. For example, home adaptations will not routinely produce savings in home-care expenditure (because the vast majority of people receiving these do not require home-care and others are so frail that adaptations cannot remove the need for care) but savings are still likely to be found through the prevention of falls or by deferring admission into residential care.

Overall, the evidence from the research powerfully illustrates that home adaptations can save health and social care expenditure under four main headings.

Reducing or removing completely an existing outlay (costs of residential care and the costs of intensive home-care)

The report's conclusions include the following:

- significant savings are found in relation to those younger disabled people who receive more intensive support packages
- the provision of adaptation and equipment that enables someone to move out of residential care produces direct savings, normally within the first year. These savings are especially large where the state is bearing the full cost of residential care
- adaptations that remove or reduce the need for daily visits pay for themselves in a time-span ranging from a few months to three years, and then produce annual savings. In the cases reviewed, annual savings varied from £1,200 to £29,000 a year
- the benefits of adaptations provided in relation to home-care for older people are often indirect. They include both the provision of assistance to existing unpaid carers and the prevention of accidents that would lead to a need for home-care; both of these may prevent or delay a move to residential care and meanwhile improve the quality of life for the older person and carer.

Saving through prevention of an outlay that would otherwise have been incurred (prevention of accidents and their associated costs, prevention of hospital admissions or move to residential care and prevention of the need for other medical treatment)

The report's conclusions include the following:

- improved lighting reduces the risk of falls
- suitable adaptations for people with visual impairments can produce health care savings
- the average cost to the state of dealing with a hip fracture is almost five times the average cost of a major home adaptation

- for parent care-givers, there is a 90% chance of musculoskeletal damage, an increased risk of falls and stress when space is inadequate and without the necessary home adaptations in place – all of which increase the cost to the NHS
- lack of timely provision of equipment may lead to additional long-term health problems for disabled people such as pressure sores, ulcers, burns and pains
- poor accessibility in the home worsens mental health.

Saving through the prevention of waste, for example underfunding causing delays in provision

The report's conclusions include the following:

- delay in adaptation provision was leading to more costly options. One person needed 4.5 additional home-care hours a week for 32 weeks, at a total cost of £1,440, after a door-widening adaptation costing £300 was delayed for 7 months for lack of funding
- delays can also result in a waste of human potential. Both housing adaptations and assistive technology have helped people into employment who would otherwise not have achieved this.

Saving through achieving better outcomes for the same expenditure

The report's conclusions include the following:

- the average cost of a DFG (£6,000) pays for a stairlift and level-access shower, a common package for older applicants. These items will last at least 5 years. The same expenditure would be enough to purchase the average home-care package (6.5 hours per week) for just 1 year and 3 months
- for the average older applicant, an adaptation package will normally pay for itself within the life-expectancy of the person concerned and will produce better value for money in terms of improved outcomes for the applicant, for example in terms of dignity and autonomy
- a bedroom extension with en-suite costs an average £32,000. This compares with the average cost of residential care of from £26,000 to £100,000 per year. Research shows the adaptations in many cases also offer a better quality of life in terms of independence and autonomy compared with residential care
- adaptations commonly produce improved quality of life for around 90% of recipients.

Zokaei K et al (2010). Report for the Wales Audit Office. Lean and systems thinking in the public sector in Wales. Lean Enterprise Research Centre

This research was commissioned by the Wales Audit Office, and focuses on efficiency in the evaluation of systems thinking in the public sector. Three case studies from local authorities are used, including the DFG service in Neath Port Talbot. In providing an overview of the improvement process carried out in the DFG service in the Borough Council, the research suggested that a timely DFG service which provided an early response could delay entry into residential care by an average of 4 years.

This would provide a cost saving of around £72,000 per DFG case (based on residential care costs of £19,760 per year, and the average cost of a DFG of £7,000). Each year's delay in providing a DFG therefore costs social care services around £18,000 per case.

These findings provided a potential and hypothetical saving with the figures calculated in retrospect rather than showing actual savings.

Pleace N (2011). The costs and benefits of preventative support services for older people. University of York

This research was commissioned by the Scottish government to review the evidence on the cost-effectiveness of preventative support services that assist older people with care and support needs to remain in their own homes. The costs of these preventative support services (handyperson schemes, home adaptations, alarm systems and telecare systems and Supporting People services) are compared with the costs of specialist housing options, such as sheltered and extra-care housing, and also with the costs of health services, as part of reviewing the value for money of preventative support services.

In terms of the cost-effectiveness of telecare, it cites previous conclusions of research evaluations, that is, the lack of conclusive evidence. It then examines the Newhaven Research (2010) concluding that it was problematic to discuss typical net costs of telecare because of:

- the wide range of telecare services with varying cost implications
- telecare services are often not used in isolation, but are often combined with other health or personal social services
- it was difficult to evaluate the costs beyond identifying that there was a mix of needs and variation in the extent of the costs.

The research by Pleace cited case studies from *Exploring the cost implications of telecare service provision* (Newhaven Research, 2010). These provide details of the full care packages with home adaptations and telecare technology together with the alternative health and social care packages which would be required without interventions. The case studies cannot distinguish between the relative contribution of telecare in isolation because of the higher care needs of the older people who used the technology, but they demonstrate how a suitable package of home adaptations, technology and home-care can potentially reduce health and social care expenditure.

Other research into telecare programmes in Scotland is also reviewed and although the author mentions the limitations of these, Pleace considers that there is evidence to indicate that telecare services, which are normally used as part of broader support packages, can significantly delay the point of admission to hospital or residential care and thus have positive economic outcomes.

Pleace cites similar conclusions with regards to home adaptations, namely, that these are sometimes used as part of an integrated support package for those with higher needs, so it is often difficult to determine the economic benefits of them, particularly where a person's physical and mental health deteriorate to a point where extra-care housing or residential care is required. Nonetheless there is evidence that home adaptations, like telecare, are most cost effective prior to extra care being required.

Clarke A (2011). Cost effectiveness of lighting adaptations. A report for the Pocklington Trust. Cambridge Centre for Housing and Planning Research

The aim of this study was to estimate the costs and benefits to the taxpayer of fitting lighting adaptations in the homes of older people at risk of falling, using data collected on the costs of lighting adaptations from a pilot scheme undertaken by the Pocklington Trust in 2007. The focus of the research was people living in their own homes. A literature review was undertaken to estimate the likelihood of lighting adaptations preventing a fall, the costs of falls to the NHS, and other costs and benefits associated with fitting lighting adaptations.

This literature review found that falls were more likely to happen in poor light, that the visually impaired were more likely to fall, and that falls prevention programmes offering a variety of home improvements (including lighting) could achieve a measurable reduction in falls of between 6 and 33%. There was, however, a lack of robust evidence on the effectiveness of lighting adaptations in reducing falling, as this has not been the main focus of previous studies.

The research then developed an analytical framework in order to establish the cost-effectiveness of lighting adaptations in the homes of partially sighted people, and a toolkit was developed for use by the Pocklington Trust or for others to update with new data in the future.

The results of the analysis were not strongly conclusive in terms of the cost-effectiveness of lighting adaptations, and the author examines the reasons for this. Some key findings are given below:

- cost per beneficiary of lighting adaptations = £1,024
- estimated costs to taxpayer per fall = £524
- financial gain to taxpayer per lighting adaptation per year = £56
- number of years lighting adaptations will last = 25
- overall saving over life of lighting adaptation=£111
- overall saving per year per adaptation = £4
- number of years to recoup spend= 23.

Snell T, Fernandez J L and Forder J (2012). Building a business case for investing in adaptive technologies in England. PSSRU discussion paper. London School of Economics

Findings from a literature review were used to build simulation models of the outcomes associated with common home adaptations at a national economic level and how changes in investment may impact on these. The research findings provided the first estimate of the outcomes of home adaptations at national (England) level. The literature informed the modelling by providing data on the costs, effectiveness (mainly in terms of falls prevention) and outcomes associated with aids and adaptations. Owing to varied findings of the review, three modelling scenarios were undertaken including the central scenario, based upon best-evidenced assumptions and a conservative estimate of potential benefits where sufficient evidence did not exist.

The costs of providing home adaptations were based on the average annual costs of adaptations cited in the literature review, the majority of which focus on major interventions that would be provided through DFGs. This therefore allowed some comparability of findings. The model did not attempt to differentiate between different types of intervention and so it was assumed that interventions cost approximately £6,000 on average and last for approximately 6 years, equating to an average annual cost of £1,000 including installation and maintenance.

Using the central scenario, the research suggests that home adaptations lead to reductions in the demand for other health and social care services worth on average £579 per recipient per annum (including both state and private costs). In addition estimated improvements in the quality of life of the dependent person are worth £1,522 per annum.

Given a client base of 45,000 individuals receiving DFG interventions (at a total cost of approximately £270 million, broadly equivalent to the total annual DFG expenditure), reductions in the demand for health and social care services are worth £156 million over the estimated lifetime of the equipment, and achieve quality of life gains of £411 million over the same period.

The authors of the discussion paper consider the biggest single limitation when modelling the economic benefits of investment in adaptive technologies to be the dearth of quality scalable evidence. They feel that evidence that adaptive technologies can pay their way is abundant, evidence base around fall prevention is reasonably strong, but indicators of wider benefits achieved with an average case are scarce.

The assumptions of the model are based on the characteristics of people eligible for state-funded adaptations in terms of physical dependency, and can be used to estimate the effects of limited increases or decreases in state provision.

Consequently, the authors indicate that their modelling would not be appropriate to assess the benefits of equipment and adaptations provided to a substantially different proportion of the population.

The King's Fund (2013). Exploring the system-wide costs of falls in older people in Torbay. The King's Fund

This research used Torbay's patient-level linked dataset to explore the cost of the health and social care pathway for older people (421 cases) admitted to hospital as a result of a fall by tracking their health and social care costs in the 12 months before and after their fall. The analysis aimed to provide a baseline against which the impact of further policy and practice changes in the area could be assessed. The key findings, which support the case for an integrated response for frail older people at risk of falls, were:

• the cost of hospital, community and social care services for patients who fell were, on average, roughly four times as much in the 12 months after hospital admission, as the costs of the admission itself
• over the 12 months that followed admission for falls, costs were 70% higher than in the 12 months before the fall
• comparing the 12 months before and after a fall, the most dramatic increase was in community care costs (160%), compared to a 37% increase in social care costs and a 35% increase in acute hospital care costs.

Keall M D et al. Home modifications to reduce injuries from falls in the Home Injury Prevention Intervention (HIPI) study: A cluster and randomised controlled trial. The Lancet, Early Online Publication, 23 September 2014. Available at: http://doi:10.1016/S01406736(14)61006-0

This New Zealand 4-year HIPI study investigated whether a package of home adaptations could reduce the rate of injuries per person per year from falls in the home requiring medical treatment as derived from administrative data for insurance claims.

A single-blind, cluster-randomised controlled trial of households was undertaken involving some 842 households

(1,848 individuals). These households lived in owner-occupied dwellings built before 1980 and at least one household member was in receipt of state benefits or subsidies. The study dealt, in the author's view, with fairly hazardous home environments with many identifiable hazards compared with previous studies. These households were randomly split between a treatment group receiving home modifications or a control group with a 3-year wait before modifications were applied. The home modifications were not tailored to individual occupants but were designed to provide a safer home environment for all occupants. Statistical modelling adjustments were made to deal with the differences in the profile of the two groups, for example age, previous falls, ethnicity.

Compared to the control group, the rate of injuries caused by falls in the home was estimated to have been reduced by 26%. For all injuries considered most relevant to the type of intervention, the rate of injuries was estimated to have fallen by 39%.

Chiatti C and Iwarsson S. Evaluation of housing adaptation interventions: Integrating the economic perspective into occupational therapy practice. Scandinavian Journal of Occupational Therapy, 2014, 21: 313–333

This article represents a predominantly Swedish perspective on the key principles and challenges of undertaking economic evaluations of home adaptations. The purpose of the article is to contribute to the development of strategies for these evaluations so they can applied by practitioners in the field, particularly occupational therapists. It uses vignettes based on the experiences of the authors and expertise in occupational therapy to demonstrate key concepts and methods for evaluations.

The authors highlight how different national frameworks for the delivery of home adaptations, for example whether these are funded or part-funded through the public purse, increase the difficulties of cross-national comparisons of economic evaluations, although some of the issues highlighted in the article do apply to the UK.

Critically, the article highlights the need for more research on the role of the occupational therapist and on determining the nature of the relevant benefits from the intervention, namely to what extent are these benefits for the client, the health and social care sector, government as a whole, the wider community or a combination of all of these.

Davison S for Care & Repair Cymru (2012). There's no place like your own home: An evaluation of the services of Care & Repair agencies in Wales. Care & Repair Cymru. Available at: www.careandrepair.org.uk/uploads/Publications/Theres_no_place_like_home_E.pdf

This publication evaluates the impact of the varied types of services provided by Care & Repair Cymru including work associated with home adaptations and energy efficiency improvements. It includes a literature review of key and recent research relating to relevant agency services as well as case studies. With regards to home adaptations, it cites the evaluation of the Welsh Rapid Response Adaptation Programme (RRAP) involving Care & Repair Cymru working in partnership with health and local authorities to develop a speedy service to carry out minor adaptations up to a maximum cost of £350. An evaluation of the programme suggests that for every £1 spent, £7.50 is saved in health and social care budgets. The publication also cites the work by Heywood and Turner (2007) and Zokaei et al (2010).

To assess some gains in health outcomes and expenditure, it uses the BRE methodology from research into the cost of poor housing (Roys et al, 2010), in that it looks at the costs of remedying Category 1 HHSRS hazards compared with the potential treatment costs to the NHS of treating injuries from these hazards.

Biel, Hanover (Scotland) and Trust Housing Associations (2012). Social Return on Investment of Stage 3 Adaptations in Sheltered and Very Sheltered Housing. Edinburgh, Bield, Hanover (Scotland) and Trust Housing Associations. Available at: www.hanover.scot/wp-content/uploads/2015/11/SROI-Adaptations-briefing-reportfinal-Sept-2011.pdf

This report includes an evaluation of the value for money provided by home adaptations for tenants in very sheltered housing (extra-care housing) through shifting the balance of care away from residential care homes and hospitals. The research used both quantitative and qualitative data sources in Scotland, together with the Adult Social Care Outcomes toolkit, benchmarking against Personal Social Services Research Unit (PSSRU) data and the social return on investment (SROI) methodology to value broader outcomes (for example improved well-being) that are more difficult to quantify. Qualitative data included a variety of data from the three housing associations, for example expenditure data and average adaptation costs.

The research estimates that in Very Sheltered Housing settings:

- the average cost of an adaptation (£2,800) has a potential £7,500 saving through reduced need for publicly-funded care
- on average each adaptation saves the Scottish health and social care system over £10,000
- £1.4 million invested in adaptations across the three housing associations in the study creates approximately £5.3 million in cost savings to the Scottish government, and a further £3.1 million in social and economic value for tenants.

On a smaller scale:

- adaptations provide a total return on investment of £5.50 to £6.00 for every £1 invested, and the Scottish government recoups £3.50 to £4.00 for every £1 it invests.

A comparison of well-being indicators such as autonomy and control, maintaining independence and social relationships, is also provided to demonstrate the benefits of tenants remaining in very sheltered housing compared with moving to care homes.

Poole T (2006). Telecare and older people. The Kings Fund

This research, written and published prior to the launch of the DoH's Whole System Demonstrator Programme, aimed to provide some background information on the key issues surrounding the introduction of telecare and related technologies. For the discussion on the cost-effectiveness of telecare provision, it references a number of existing studies/ local pilot schemes, the majority of which were small or medium size in scale and, according to the author; their findings cannot be applied at national level.

The author concluded, among other things, that the evidence base for both the impact and cost-effectiveness of telecare was limited at the time due to the lack of rigorous data, with quality of life improvements being easier to demonstrate. However, she felt that there was a general consensus that even low-level telecare could reduce the demand for care home and hospital beds but qualified this by highlighting the lack of a consistent

framework for these costs assessments. The research also said that the beneficial effect of telecare was likely to be most effective for people in the low and medium frailty groups, who are less likely to be eligible for funding for adaptations.

One of the key difficulties in undertaking evaluations was considered to be the organisational and structural conditions under which telecare is administered, namely, the practical challenges of running pilot studies over a long period of time, retaining sufficient participants and multiple agencies for a number of years.

Davies A and Newman S (2011). Evaluating telecare and telehealth interventions. WSDAN briefing paper. The King's Fund

This paper's primary aim is to provide guidance on how to carry out good quality evaluations of telecare and telehealth interventions. The authors identify different types of previous evaluations, and briefly review the evidence from systematic reviews of telecare and telehealth to illustrate key issues they feel need to be addressed in future evaluations.

Their review of existing evaluations in many ways mirrors the findings of Poole in Telecare and older people (2006), in terms of citing the need for more and better quality evidence on the economic outcomes of telecare and telehealth provision. In particular they cite the lack of cost-benefit analysis for projects after a 12-month period.

Steventon A et al. Effect of telecare on use of health and social care services: Findings from the Whole Systems Demonstrator cluster randomised trial. Age and Ageing, 2013, 42 (4) 501–80. Available at: http://doi: 10.1093/ageing/aft008

Research by the academic institutions involved in the evaluation of the Whole Systems Demonstrator Trial to assess the impact of telecare on the use of social and health care.

The study involved 2,600 people with social care needs from 217 general practices in three areas in England. The trial compared telecare with usual care, using randomised cluster sampling of patients from the general practices. Participants were followed up for 12 months and analyses were conducted as intention-to-treat. The trial data were linked to personal data on care that was funded at least in part by local authorities or the NHS. The principal aim was to measure the proportion of people admitted to hospital within 12 months.

Secondary analysis was undertaken in relation to mortality, rates of secondary care use, contacts with general practitioners and practice nurses, proportion of people admitted to permanent residential or nursing care, weeks in domiciliary social care and notional costs.

The research showed that 46.8% of intervention participants (with telecare) were admitted to hospital, compared with 49.2% of other patients. The differences were not found to be statistically significant, except when adjusted for characteristics such age, sex, area and health conditions. The results were also not statistically significant when compared to a predictive model of hospital admissions (based on general practice and hospital data). Secondary analysis including impacts on social care use also showed no statistical significant difference between the two patient groups. The authors concluded that telecare, as implemented in the Whole Systems Demonstrator trial, did not lead to significant reductions in service use, at least in terms of results assessed over 12 months.

References

Adams S (2015). Disabled Facilities Grants: A system of help with home adaptations for disabled people in England – an overview. Nottingham, Care & Repair England. Available at: http:// careandrepair-england.org. uk/wp-content/uploads/2014/12/Provision-of-help-home-adaptations-for-disabled-people-inEngland-briefing-July-2015.pdf.

Age UK (2014). Housing in later life. London, Age UK.

Ambrose P (2001). Living conditions and health promotion strategies. Journal of the Royal Society for the Promotion of Health, 121.1 pp 9–15.

Astral Advisory (2013). Disabled Facilities Grants in England: A research report. Glossop, Astral Advisory.

Bield, Hanover (Scotland) and Trust Housing Associations. Spend now. Save for the future. A social return on investment study of adaptations. Available at: www.trustha.org.uk/media/uploads/ documents/SROI_ Adaptations_-_spend_now_save_for_the_ future.pdf.

Bield, Hanover (Scotland) and Trust Housing Associations (2012). Measuring the social return on investment of stage 3 adaptations and Very Sheltered Housing in Scotland. Edinburgh, Bield, Hanover (Scotland) and Trust Housing Associations. Available at: www.hanover. scot/wp-content/uploads/2015/11/ SROI-Adaptations-briefing-report-final-Sept-2011.pdf.

BRE for the Department for Communities and Local Government (DCLG) (2011). Disabled Facilities Grant allocation methodology and means test. Final report. London, DCLG.

Bristol University for the Office of the Deputy Prime Minister (ODPM) (2005). Reviewing the disabled facilities grant programme. London, ODPM.

Chiatti C and Iwarsson S. Evaluation of housing adaptation interventions: Integrating the economic perspective into occupational therapy practice. Scandinavian Journal of Occupational Therapy, 2014, 21: 313–333.

Clarke A (2011). Cost effectiveness of lighting adaptations. A report for the Pocklington Trust. Cambridge, Cambridge Centre for Housing and Planning Research.

Clark M and Goodwin N (2010). Sustaining innovation in telehealth and telecare WSDAN briefing paper. London, The Kings Fund. Available at: www.kingsfund.org.uk/sites/files/kf/Sustaininginnovation-telehealth-telecare-wsdan-mike-clark-nick-goodwinoctober-2010.pdf.

Davies A and Newman S (2011). Evaluating telecare and telehealth interventions. WSDAN briefing paper. London, The King's Fund.

Davison S for Care & Repair Cymru (2012). There's no place like your own home. An evaluation of the services of Care & Repair agencies in Wales. Cardiff, Care & Repair Cymru. Available at: www.careandrepair. org.uk/uploads/Publications/Theres_no_ place_like_home_E.pdf.

Department for Communities and Local Government (DCLG) (2011). Laying the foundations: A housing strategy for England. London, DCLG.

DCLG (2012). National evaluation of the Handyperson Programme. London, DCLG.

Department of Health (DoH) (2012). Improving outcomes and supporting transparency, Part 1: A public health outcomes framework for England, 2013–2016. Headline findings available at: www.gov.uk/government/uploads/system/uploads/attachment_ data/file/215264/dh_131689.pdf.

DoH (2013). Research and development work relating to assistive technology, 2012–13. Presented to Parliament pursuant to Section 22 of the Chronically Sick and Disabled Persons Act 1970. London, DoH. Available at: www.gov.uk/government/uploads/system/ uploads/ attachment_data/file/336321/2012-13.pdf.

DoH (2014). Care and support statutory guidance, issued under the Care Act 2014. London, DoH.

DoH and DCLG (2016). 2016/17 Better Care Fund: Policy framework. London, DoH and DCLG. Available at: www.gov.uk/government/uploads/system/uploads/attachment_data/file/490559/BCF_ Policy_ Framework_2016-17.pdf.

Foundations (2010). Adapting for a lifetime: The key role of home improvement agencies in adaptations delivery. Glossop, Foundations.

Garrett H, Davidson M, Roys M, Nicol S and Mason V (2014). Quantifying the health benefits of the Decent Homes programme. BRE FB 64. Bracknell, IHS BRE Press.

Heywood F (2001). Money well spent: The effectiveness and value of housing adaptations. Bristol, The Policy Press and the Joseph Rowntree Foundation.

Heywood F and Turner L (2007). Better outcomes, lower costs: Implications for health and social care budgets of investments in housing adaptations, improvements and equipment – a review of the evidence. London, Office for Disability Issues, Department for Work and Pensions.

Home Adaptations Consortium (2013). A detailed guide to related legislation, guidance and good practice. Nottingham, Care & Repair England.

Jarrett T (2012). The Supporting People programme. House of Commons Library, Research Paper 12/40, 16 July.

Keall M D et al. Home modifications to reduce injuries from falls in the Home Injury Prevention Intervention (HIPI) study: A cluster and randomised controlled trial. The Lancet, Early Online Publication, 23 September 2014. Available at: http://doi:10.1016/ S01406736(14)61006-0.

The King's Fund (2013). Exploring the system-wide costs of falls in older people in Torbay. London, The Kings Fund.

Newhaven Research (2010). Exploring the cost implications of telecare service provision. Edinburgh, Newhaven Research.

Nicol S, Roys M and Garret H (2015). Briefing paper. The cost of poor housing to the NHS. Available at: www.bre.co.uk/page.jsp?id=3611.

Nicol S, Roys M, Davidson M, Summers C, Ormandy D and Ambrose P (2010). Quantifying the cost of poor housing. BRE IP 16/10. Bracknell, IHS BRE Press.

Office for National Statistics (2013). National population projections, 2012-based statistical bulletin. London, Office for National Statistics. Available at: www.ons.gov.uk/ons/rel/npp/nationalpopulation-projections/2012-based-projections/stb-2012-basednpp-principal-and-key-variants.html#tab-Key-Points.

Papworth Trust (2012). Home solutions to our care crisis. Cambridge, Papworth Trust.

Pleace N (2011). The costs and benefits of preventative support services for older people. York, University of York.

Poole T (2006). Telecare and older people. London, The Kings Fund.

Roys M, Davidson M, Nicol S, Ormandy D and Ambrose P (2010). The real cost of poor housing. BRE FB 23. Bracknell, IHS BRE Press.

Snell T, Fernandez J L and Forder J (2012). Building a business case for investing in adaptive technologies in England. PSSRU Discussion Paper. London, London School of Economics.

Steventon A et al. Effect of telecare on use of health and social care services: Findings from the Whole Systems Demonstrator cluster randomised trial. Age and Ageing, 2013, 42 (4) 501–80. Available at: http://doi: 10.1093/ageing/aft008.

Wilson W (2013). Disabled Facilities Grants (England). House of Commons Library standard note: SN/SP/3011. Last updated: 18 December 2013. Available at: http://researchbriefings.files.parliament.uk/documents/SN03011/SN03011.pdf.

Zokaei K et al (2010). Report for the Wales Audit Office. Lean and systems thinking in the public sector in Wales. Cardiff, Lean Enterprise Research Centre.

Publications from IHS BRE Press

External fire spread. 2nd edn. **BR 187**

Site layout planning for daylight and sunlight. 2nd edn. **BR 209**

Radon: guidance on protective measures for new buildings. 2015 edn. **BR 211**

Conventions for calculating linear thermal transmittance and temperature factors. 2nd edn. **BR 497**

Fire safety and security in retail premises. **BR 508**

Automatic fire detection and alarm systems. **BR 510**

Handbook for the structural assessment of large panel system (LPS) dwelling blocks for accidental loading. **BR 511**

Integrating BREEAM throughout the design process: a guide to achieving higher BREEAM and Code for Sustainable Homes ratings through incorporation with the RIBA Outline Plan of Work and other procurement routes. **FB 28**

Design fires for use in fire safety engineering. **FB 29**

Ventilation for healthy buildings: reducing the impact of urban pollution. **FB 30**

Financing UK carbon reduction projects. **FB 31**

The cost of poor housing in Wales. **FB 32**

Dynamic comfort criteria for structures: a review of UK standards, codes and advisory documents. **FB 33**

Water mist fire protection in offices: experimental testing and development of a test protocol. **FB 34**

Airtightness in commercial and public buildings. 3rd edn. FB 35

Biomass energy. **FB 36**

Environmental impact of insulation. **FB 37**

Environmental impact of vertical cladding. **FB 38**

Environmental impact of floor finishes: incorporating The Green Guide ratings for floor finishes. **FB 39**

LED lighting. **FB 40**

Radon in the workplace. 2nd edn. **FB 41**

U-value conventions in practice. **FB 42**

Lessons learned from community-based microgeneration projects: the impact of renewable energy capital grant schemes. **FB 43**

Energy management in the built environment: a review of best practice. **FB 44**

The cost of poor housing in Northern Ireland. **FB 45**

Ninety years of housing, 1921–2011. **FB 46**

BREEAM and the Code for Sustainable Homes on the London 2012 Olympic Park. **FB 47**

Saving money, resources and carbon through SMARTWaste. **FB 48**

Concrete usage in the London 2012 Olympic Park and the

Olympic and Paralympic Village and its embodied carbon content. **FB 49**

A guide to the use of urban timber. **FB 50**

Low flow water fittings: will people accept them? **FB 51**

Evacuating vulnerable and dependent people from buildings in an emergency. **FB 52**

Refurbishing stairs in dwellings to reduce the risk of falls and injuries. **FB 53**

Dealing with difficult demolition wastes. **FB 54**

Security glazing: is it all that it's cracked up to be? **FB 55**

The essential guide to retail lighting. **FB 56**

Environmental impact of metals. **FB 57**

Environmental impact of brick, stone and concrete. **FB 58**

Design of low-temperature domestic heating systems. **FB 59**

Performance of photovoltaic systems on non-domestic buildings. **FB 60**

Reducing thermal bridging at junctions when designing and installing solid wall insulation. **FB 61**

Housing in the UK. **FB 62**

Delivering sustainable buildings. **FB 63**

Quantifying the health benefits of the Decent Homes programme. **FB 64**

The cost of poor housing in London. **FB 65**

Environmental impact of windows. **FB 66**

Environmental impact of biomaterials and biomass. **FB 67**

DC isolators for photovoltaic systems. **FB 68**

Computational fluid dynamics in building design. **FB 69**

Design of durable concrete structures. **FB 70**

The age and construction of English homes. **FB 71**

A technical guide to district heating. **FB 72**

Changing energy behaviour in the workplace. **FB 73**

Lighting and health. **FB 74**

Building on fill: geotechnical aspects. 3rd edn. **FB 75**

Changing patterns in domestic energy use. **FB 76**

Embedded security: procuring an effective facility protective security system. **FB 77**

Performance of exemplar buildings in use: bridging the performance gap. **FB 78**

Designing out unintended consequences when applying solid wall insulation. **FB 79**

Applying Fabric First principles. **FB 80**

For a complete list of IHS BRE Press publications visit www.brebookshop.com